Chemistry Workbook

FOR DUMMIES®

A Wiley Brand

2nd Edition

by Peter J. Mikulecky, PhD,
and Christopher Hren

FOR DUMMIES®
A Wiley Brand

Chemistry Workbook For Dummies,® 2nd Edition

Published by: **John Wiley & Sons, Inc.,** 111 River Street, Hoboken, NJ 07030-5774, www.wiley.com

Copyright © 2015 by John Wiley & Sons, Inc., Hoboken, New Jersey

Published simultaneously in Canada

No part of this publication may be reproduced, stored in a retrieval system or transmitted in any form or by any means, electronic, mechanical, photocopying, recording, scanning or otherwise, except as permitted under Sections 107 or 108 of the 1976 United States Copyright Act, without the prior written permission of the Publisher. Requests to the Publisher for permission should be addressed to the Permissions Department, John Wiley & Sons, Inc., 111 River Street, Hoboken, NJ 07030, (201) 748-6011, fax (201) 748-6008, or online at http://www.wiley.com/go/permissions.

Trademarks: Wiley, For Dummies, the Dummies Man logo, Dummies.com, Making Everything Easier, and related trade dress are trademarks or registered trademarks of John Wiley & Sons, Inc., and may not be used without written permission. All other trademarks are the property of their respective owners. John Wiley & Sons, Inc., is not associated with any product or vendor mentioned in this book.

LIMIT OF LIABILITY/DISCLAIMER OF WARRANTY: WHILE THE PUBLISHER AND AUTHOR HAVE USED THEIR BEST EFFORTS IN PREPARING THIS BOOK, THEY MAKE NO REPRESENTATIONS OR WARRANTIES WITH RESPECT TO THE ACCURACY OR COMPLETENESS OF THE CONTENTS OF THIS BOOK AND SPECIFICALLY DISCLAIM ANY IMPLIED WARRANTIES OF MERCHANTABILITY OR FITNESS FOR A PARTICULAR PURPOSE. NO WARRANTY MAY BE CREATED OR EXTENDED BY SALES REPRESENTATIVES OR WRITTEN SALES MATERIALS. THE ADVICE AND STRATEGIES CONTAINED HEREIN MAY NOT BE SUITABLE FOR YOUR SITUATION. YOU SHOULD CONSULT WITH A PROFESSIONAL WHERE APPROPRIATE. NEITHER THE PUBLISHER NOR THE AUTHOR SHALL BE LIABLE FOR DAMAGES ARISING HEREFROM.

For general information on our other products and services, please contact our Customer Care Department within the U.S. at 877-762-2974, outside the U.S. at 317-572-3993, or fax 317-572-4002. For technical support, please visit www.wiley.com/techsupport.

Wiley publishes in a variety of print and electronic formats and by print-on-demand. Some material included with standard print versions of this book may not be included in e-books or in print-on-demand. If this book refers to media such as a CD or DVD that is not included in the version you purchased, you may download this material at http://booksupport.wiley.com/. For more information about Wiley products, visit www.wiley.com.

Library of Congress Control Number: 2014908772

ISBN 978-1-118-94004-4 (pbk); ISBN 978-1-118-94005-1 (ebk); ISBN 978-1-118-94006-8 (ebk)

Manufactured in the United States of America

10 9 8 7 6 5 4 3 2 1

Contents at a Glance

Table of Contents

Introduction

. . . .

"The first essential in chemistry is that you should perform practical work and conduct experiments, for he who performs not practical work nor makes experiments will never attain the least degree of mastery."

—Jābir ibn Hayyān, 8th century

"One of the wonders of this world is that objects so small can have such consequences: Any visible lump of matter — even the merest speck — contains more atoms than there are stars in our galaxy."

—Peter W. Atkins, 20th century

Chemistry is at once practical and wondrous, humble and majestic. And for someone studying it for the first time, chemistry can be tricky.

That's why we wrote this book. Chemistry is wondrous. Workbooks are practical. Practice makes perfect. This chemistry workbook will help you practice many types of chemistry problems with the solutions nicely laid out.

About This Book

When you're fixed in the thickets of stoichiometry or bogged down by buffered solutions, you've got little use for rapturous poetry about the atomic splendor of the universe. What you need is a little practical assistance. Subject by subject, problem by problem, this book extends a helping hand to pull you out of the thickets and bogs.

The topics covered in this book are the topics most often covered in a first-year chemistry course in high school or college. The focus is on problems — problems like the ones you may encounter in homework or on exams. We give you just enough theory to grasp the principles at work in the problems. Then we tackle example problems. Then *you* tackle practice problems. The best way to succeed at chemistry is to practice. Practice more. And then practice even more. Watching your teacher do the problems or reading about them isn't enough. Michael Jordan didn't develop a jump shot by watching other people shoot a basketball. He practiced. A lot. Using this workbook, you can, too (but chemistry, not basketball).

This workbook is modular. You can pick and choose those chapters and types of problems that challenge you the most; you don't have to read this book cover to cover if you don't want to. If certain topics require you to know other topics in advance, we tell you so and point you in the right direction. Most importantly, we provide a worked-out solution and explanation for every problem.

Foolish Assumptions

We assume you have a basic facility with algebra and arithmetic. You should know how to solve simple equations for an unknown variable. You should know how to work with exponents and logarithms. That's about it for the math. At no point do we ask you to, say, consider the contradictions between the Schrödinger equation and stochastic wavefunction collapse.

We assume you're a high school or college student and have access to a secondary school–level (or higher) textbook in chemistry or some other basic primer, such as *Chemistry For Dummies,* 2nd Edition (written by John T. Moore, EdD, and published by Wiley). We present enough theory in this workbook for you to tackle the problems, but you'll benefit from a broader description of basic chemical concepts. That way, you'll more clearly understand how the various pieces of chemistry operate within a larger whole — you'll see the compound for the elements, so to speak.

We assume you don't like to waste time. Neither do we. Chemists in general aren't too fond of time-wasting, so if you're impatient for progress, you're already part-chemist at heart.

Icons Used in This Book

You'll find a selection of helpful icons nicely nestled along the margins of this workbook. Think of them as landmarks, familiar signposts to guide you as you cruise the highways of chemistry.

Within already pithy summaries of chemical concepts, passages marked by this icon represent the pithiest must-know bits of information. You'll need to know this stuff to solve problems.

Sometimes there's an easy way and a hard way. This icon alerts you to passages intended to highlight an easier way. It's worth your while to linger for a moment. You may find yourself nodding quietly as you jot down a grateful note or two.

Chemistry may be a practical science, but it also has its pitfalls. This icon raises a red flag to direct your attention to easily made errors or other tricky items. Pay attention to this material to save yourself from needless frustration.

Within each section of a chapter, this icon announces, "Here ends theory" and "Let the practice begin." Alongside the icon is an example problem that employs the very concept covered in that section. An answer and explanation accompany each practice problem.

Beyond the Book

In addition to the topics we cover in this book, you can find even more information online. The Cheat Sheet at www.dummies.com/cheatsheet/chemistrywb provides some quick and useful tips for solving the most common types of chemistry problems you'll see. If you'd like to pick up some chemistry-specific study tips, find out more about solid-state chemistry, or see a valuable alternative to determining concentration in molarity, go to www.dummies.com/extras/chemistrywb.

Where to Go from Here

Where you go from here depends on your situation and your learning style:

✔ If you're currently enrolled in a chemistry course, you may want to scan the table of contents to determine what material you've already covered in class and what you're covering right now. Use this book as a supplement to clarify things you don't understand or to practice concepts that you're struggling with.

✔ If you're brushing up on forgotten chemistry, scan the chapters for example problems. As you read through them, you'll probably have one of two responses: 1) "Ahhh, yes . . . I remember that" or 2) "Oooh, no . . . I so do *not* remember that." Let your responses guide you.

✔ If you're just beginning a chemistry course, you can follow along in this workbook, using the practice problems to supplement your homework or as extra pre-exam practice. Alternatively, you can use this workbook to preview material before you cover it in class, sort of like a spoonful of sugar to help the medicine go down.

✔ If you bought this book a week before your final exam and are just now trying to figure out what this whole "chemistry" thing is about, well, good luck. The best way to start in that case is to determine what exactly is going to be on your exam and to study only those parts of this book. Due to time constraints or the proclivities of individual teachers/ professors, not everything is covered in every chemistry class.

No matter the reason you have this book in your hands now, there are three simple steps to remember:

1. **Don't just read it; do the practice problems.**

2. **Don't panic.**

3. **Do more practice problems.**

Anyone can do chemistry given enough desire, focus, and time. Keep at it, and you'll get an element on the periodic table named after you soon enough.

Part I

Getting Cozy with Numbers, Atoms, and Elements

Visit www.dummies.com for great (and free!) Dummies content online.

In this part . . .

✔ Discover how to deal with, organize, and use all the numbers that play a huge role in chemistry. In particular, find out about exponential and scientific notation as well as precision and accuracy.

✔ Convert many types of units that exist across the scientific world. From millimeters to kilometers and back again, you find conversions here.

✔ Determine the arrangement and structure of subatomic particles in atoms. Protons, neutrons, and electrons play a central role in everything chemistry, and you find their most basic properties in this part.

✔ Get the scoop on the arrangement of the periodic table and the properties it conveys for each group of elements. Just from looking at the periodic table and its placement of elements, you can find so much information, from electron energy levels to ionic charge and more.

Chapter 1

Noting Numbers Scientifically

● ●

In This Chapter

▶ Crunching numbers in scientific and exponential notation

▶ Telling the difference between accuracy and precision

▶ Doing math with significant figures

● ●

*L*ike any other kind of scientist, a chemist tests hypotheses by doing experiments. Better tests require more reliable measurements, and better measurements are those that have more accuracy and precision. This explains why chemists get so giggly and twitchy about high-tech instruments: Those instruments take better measurements!

How do chemists report their precious measurements? What's the difference between accuracy and precision? And how do chemists do math with measurements? These questions may not keep you awake at night, but knowing the answers to them will keep you from making rookie mistakes in chemistry.

Using Exponential and Scientific Notation to Report Measurements

Because chemistry concerns itself with ridiculously tiny things like atoms and molecules, chemists often find themselves dealing with extraordinarily small or extraordinarily large numbers. Numbers describing the distance between two atoms joined by a bond, for example, run in the ten-billionths of a meter. Numbers describing how many water molecules populate a drop of water run into the trillions of trillions.

To make working with such extreme numbers easier, chemists turn to scientific notation, which is a special kind of exponential notation. *Exponential notation* simply means writing a number in a way that includes exponents. In scientific notation, every number is written as the product of two numbers, a coefficient and a power of 10. In plain old exponential notation, a coefficient can be any value of a number multiplied by a power with a base of 10 (such as 10^4). But scientists have rules for coefficients in scientific notation. In *scientific notation,* the coefficient is always at least 1 and always less than 10. For example, the coefficient could be 7, 3.48, or 6.0001.

To convert a very large or very small number to scientific notation, move the decimal point so it falls between the first and second digits. Count how many places you moved the decimal point to the right or left, and that's the power of 10. If you moved the decimal point to the left, the exponent on the 10 is positive; to the right, it's negative. (Here's another easy way to remember the sign on the exponent: If the initial number value is greater than 1, the exponent will be positive; if the initial number value is between 0 and 1, the exponent will be negative.)

To convert a number written in scientific notation back into decimal form, just multiply the coefficient by the accompanying power of 10.

Q. Convert 47,000 to scientific notation.

A. $47,000 = 4.7 \times 10^4$. First, imagine the number as a decimal:

$$47,000.$$

Next, move the decimal point so it comes between the first two digits:

$$4.7000$$

Then count how many places to the left you moved the decimal (four, in this case) and write that as a power of 10: 4.7×10^4.

Q. Convert 0.007345 to scientific notation.

A. $0.007345 = 7.345 \times 10^{-3}$. First, put the decimal point between the first two non-zero digits:

$$7.345$$

Then count how many places to the right you moved the decimal (three, in this case) and write that as a power of 10: $0.007345 = 7.345 \times 10^{-3}$.

---

1. Convert 200,000 into scientific notation.

Solve It

2. Convert 80,736 into scientific notation.

Solve It

---

3. Convert 0.00002 into scientific notation.

Solve It

4. Convert 6.903×10^2 from scientific notation into decimal form.

Solve It

Multiplying and Dividing in Scientific Notation

A major benefit of presenting numbers in scientific notation is that it simplifies common arithmetic operations. The simplifying abilities of scientific notation are most evident in multiplication and division. (As we note in the next section, addition and subtraction benefit from exponential notation but not necessarily from strict scientific notation.)

To multiply two numbers written in scientific notation, multiply the coefficients and then add the exponents. To divide two numbers, simply divide the coefficients and then subtract the exponent of the *denominator* (the bottom number) from the exponent of the *numerator* (the top number).

Q. Multiply using the shortcuts of scientific notation: $(1.4 \times 10^2) \times (2.0 \times 10^{-5})$.

A. $\mathbf{2.8 \times 10^{-3}}$. First, multiply the coefficients:

$$1.4 \times 2.0 = 2.8$$

Next, add the exponents of the powers of 10:

$$10^2 \times 10^{-5} = 10^{2+(-5)} = 10^{-3}$$

Finally, join your new coefficient to your new power of 10:

$$2.8 \times 10^{-3}$$

Q. Divide using the shortcuts of scientific notation: $\dfrac{3.6 \times 10^{-3}}{1.8 \times 10^4}$.

A. $\mathbf{2.0 \times 10^{-7}}$. First, divide the coefficients:

$$\frac{3.6}{1.8} = 2.0$$

Next, subtract the exponent in the denominator from the exponent in the numerator:

$$\frac{10^{-3}}{10^4} = 10^{-3-4} = 10^{-7}$$

Then join your new coefficient to your new power of 10:

$$2.0 \times 10^{-7}$$

5. Multiply $(2.2 \times 10^9) \times (5.0 \times 10^{-4})$.

Solve It

6. Divide $\dfrac{9.3 \times 10^{-5}}{3.1 \times 10^2}$.

Solve It

7. Using scientific notation, multiply 52×0.035.

Solve It

8. Using scientific notation, divide $\frac{0.00809}{20.3}$.

Solve It

Using Exponential Notation to Add and Subtract

Addition or subtraction gets easier when you express your numbers as coefficients of identical powers of 10. To wrestle your numbers into this form, you may need to use coefficients less than 1 or greater than 10. So scientific notation is a bit too strict for addition and subtraction, but exponential notation still serves you well.

To add two numbers easily by using exponential notation, first express each number as a coefficient and a power of 10, making sure that 10 is raised to the same exponent in each number. Then add the coefficients. To subtract numbers in exponential notation, follow the same steps but subtract the coefficients.

Q. Use exponential notation to add these numbers: $3{,}710 + (2.4 \times 10^2)$.

A. $\mathbf{39.5 \times 10^2}$. First, convert both numbers to the same power of 10:

$$37.1 \times 10^2 \text{ and } 2.4 \times 10^2$$

Next, add the coefficients:

$$37.1 + 2.4 = 39.5$$

Finally, join your new coefficient to the shared power of 10:

$$39.5 \times 10^2$$

Q. Use exponential notation to subtract: $0.0743 - 0.0022$.

A. $\mathbf{7.21 \times 10^{-2}}$. First, convert both numbers to the same power of 10:

$$7.43 \times 10^{-2} \text{ and } 0.22 \times 10^{-2}$$

Next, subtract the coefficients:

$$7.43 - 0.22 = 7.21$$

Then join your new coefficient to the shared power of 10:

$$7.21 \times 10^{-2}$$

9. Add $(398 \times 10^{-6}) + (147 \times 10^{-6})$.

Solve It

10. Subtract $(7.685 \times 10^{5}) - (1.283 \times 10^{5})$.

Solve It

11. Use exponential notation to add $0.00206 + 0.0381$.

Solve It

12. Use exponential notation to subtract $9,352 - 431$.

Solve It

Distinguishing between Accuracy and Precision

Accuracy and precision, precision and accuracy . . . same thing, right? Chemists everywhere gasp in horror, reflexively clutching their pocket protectors — accuracy and precision are different!

✔ **Accuracy:** Accuracy describes how closely a measurement approaches an actual, true value.

✔ **Precision:** Precision, which we discuss more in the next section, describes how close repeated measurements are to one another, regardless of how close those measurements are to the actual value. The bigger the difference between the largest and smallest values of a repeated measurement, the less precision you have.

The two most common measurements related to accuracy are *error* and *percent error:*

✔ **Error:** Error measures accuracy, the difference between a measured value and the actual value:

$$\text{Actual value} - \text{Measured value} = \text{Error}$$

✔ **Percent error:** Percent error compares error to the size of the thing being measured:

$$\frac{|\text{Error}|}{\text{Actual value}} = \text{Fraction error}$$

$$\text{Fraction error} \times 100 = \text{Percent error}$$

Being off by 1 meter isn't such a big deal when measuring the altitude of a mountain, but it's a shameful amount of error when measuring the height of an individual mountain climber.

Q. A police officer uses a radar gun to clock a passing Ferrari at 131 miles per hour (mph). The Ferrari was really speeding at 127 mph. Calculate the error in the officer's measurement.

A. **–4 mph.** First, determine which value is the actual value and which is the measured value:

• Actual value = 127 mph

• Measured value = 131 mph

Then calculate the error by subtracting the measured value from the actual value:

$$\text{Error} = 127 \text{ mph} - 131 \text{ mph} = -4 \text{ mph}$$

Q. Calculate the percent error in the officer's measurement of the Ferrari's speed.

A. **3.15%.** First, divide the error's absolute value (the size, as a positive number) by the actual value:

$$\frac{|-4 \text{ mph}|}{127 \text{ mph}} = \frac{4 \text{mph}}{127 \text{ mph}} = 0.0315$$

Next, multiply the result by 100 to obtain the percent error:

$$\text{Percent error} = 0.0315 \times 100 = 3.15\%$$

13. Two people, Reginald and Dagmar, measure their weight in the morning by using typical bathroom scales, instruments that are famously unreliable. The scale reports that Reginald weighs 237 pounds, though he actually weighs 256 pounds. Dagmar's scale reports her weight as 117 pounds, though she really weighs 129 pounds. Whose measurement incurred the greater error? Who incurred a greater percent error?

Solve It

14. Two jewelers were asked to measure the mass of a gold nugget. The true mass of the nugget is 0.856 grams (g). Each jeweler took three measurements. The average of the three measurements was reported as the "official" measurement with the following results:

• **Jeweler A:** 0.863 g, 0.869 g, 0.859 g

• **Jeweler B:** 0.875 g, 0.834 g, 0.858 g

Which jeweler's official measurement was more accurate? Which jeweler's measurements were more precise? In each case, what was the error and percent error in the official measurement?

Solve It

Expressing Precision with Significant Figures

When you know how to express your numbers in scientific notation and how to distinguish between precision and accuracy (we cover both topics earlier in this chapter), you can bask in the glory of a new skill: using scientific notation to express precision. The beauty of this system is that simply by looking at a measurement, you know just how precise that measurement is.

When you report a measurement, you should include digits only if you're really confident about their values. Including a lot of digits in a measurement means something — it means that you really know what you're talking about — so we call the included digits *significant figures*. The more significant figures (sig figs) in a measurement, the more accurate that measurement must be. The last significant figure in a measurement is the only figure that includes any uncertainty, because it's an estimate. Here are the rules for deciding what is and what isn't a significant figure:

✔ **Any nonzero digit is significant.** So 6.42 contains three significant figures.

✔ **Zeros sandwiched between nonzero digits are significant.** So 3.07 contains three significant figures.

- ✔ **Zeros on the left side of the first nonzero digit are *not* significant.** So 0.0642 and 0.00307 each contain three significant figures.

- ✔ **One or more *final zeros* (zeros that end the measurement) used after the decimal point are significant.** So 1.760 has four significant figures, and 1.7600 has five significant figures. The number 0.0001200 has only four significant figures because the first zeros are not final.

- ✔ **When a number has no decimal point, any zeros after the last nonzero digit *may* or *may not* be significant.** So in a measurement reported as 1,370, you can't be certain whether the 0 is a certain value or is merely a placeholder.

  Be a good chemist. Report your measurements in scientific notation to avoid such annoying ambiguities. (See the earlier section "Using Exponential and Scientific Notation to Report Measurements" for details on scientific notation.)

- ✔ **If a number is already written in scientific notation, then all the digits in the coefficient are significant.** So the number 3.5200×10^{-6} has five significant figures due to the five digits in the coefficient.

- ✔ **Numbers from counting (for example, 1 kangaroo, 2 kangaroos, 3 kangaroos) or from defined quantities (say, 60 seconds per 1 minute) are understood to have an unlimited number of significant figures.** In other words, these values are completely certain.

The number of significant figures you use in a reported measurement should be consistent with your certainty about that measurement. If you know your speedometer is routinely off by 5 miles per hour, then you have no business protesting to a policeman that you were going only 63.2 mph in a 60 mph zone.

Q. How many significant figures are in the following three measurements?

a. 20,175 yards

b. 1.75 yards

c. 1.750 yards

A. **a) Five, b) three, and c) four significant figures.** In the first measurement, all digits are nonzero, except for a 0 that's sandwiched between nonzero digits, which counts as significant. The coefficient in the second measurement contains only nonzero digits, so all three digits are significant. The coefficient in the third measurement contains a 0, but that 0 is the final digit and to the right of the decimal point, so it's significant.

15. Identify the number of significant figures in each measurement:

    a. 76.093×10^{-2} meters

    b. 0.000769 meters

    c. 769.3 meters

Solve It

16. In chemistry, the potential error associated with a measurement is often reported alongside the measurement, as in 793.4 ± 0.2 grams. This report indicates that all digits are certain except the last, which may be off by as much as 0.2 grams in either direction. What, then, is wrong with the following reported measurements?

    a. 893.7 ± 1 gram

    b. 342 ± 0.01 gram

Solve It

Doing Arithmetic with Significant Figures

Doing chemistry means making a lot of measurements. The point of spending a pile of money on cutting-edge instruments is to make really good, really precise measurements. After you've got yourself some measurements, you roll up your sleeves, hike up your pants, and do some math.

When doing math in chemistry, you need to follow some rules to make sure that your sums, differences, products, and quotients honestly reflect the amount of precision present in the original measurements. You can be honest (and avoid the skeptical jeers of surly chemists) by taking things one calculation at a time, following a few simple rules. One rule applies to addition and subtraction, and another rule applies to multiplication and division.

> ✔ **Adding or subtracting:** Round the sum or difference to the same number of decimal places as the measurement with the fewest decimal places. Rounding like this is honest, because you're acknowledging that your answer can't be any more precise than the least-precise measurement that went into it.

> ✔ **Multiplying or dividing:** Round the product or quotient so that it has the same number of significant figures as the least-precise measurement — the measurement with the fewest significant figures.

Notice the difference between the two rules. When you add or subtract, you assign significant figures in the answer based on the number of decimal places in each original measurement. When you multiply or divide, you assign significant figures in the answer based on the smallest number of significant figures from your original set of measurements.

Caught up in the breathless drama of arithmetic, you may sometimes perform multi-step calculations that include addition, subtraction, multiplication, and division, all in one go. No problem. Follow the normal order of operations, doing multiplication and division first, followed by addition and subtraction. At each step, follow the simple significant-figure rules, and then move on to the next step.

Q. Express the following sum with the proper number of significant figures:

$$35.7 \text{ miles} + 634.38 \text{ miles} + 0.97 \text{ miles} = ?$$

A. **671.1 miles.** Adding the three values yields a raw sum of 671.05 miles. However, the 35.7 miles measurement extends only to the tenths place. Therefore, you round the answer to the tenths place, from 671.05 to 671.1 miles.

Q. Express the following product with the proper number of significant figures:

$$27 \text{ feet} \times 13.45 \text{ feet} = ?$$

A. 3.6×10^2 **feet².** Of the two measurements, one has two significant figures (27 feet) and the other has four significant figures (13.45 feet). The answer is therefore limited to two significant figures. You need to round the raw product, 363.15 feet². You could write 360 feet², but doing so may imply that the final 0 is significant and not just a placeholder. For clarity, express the product in scientific notation, as 3.6×10^2 feet².

17. Express this difference using the appropriate number of significant figures:

$$127.379 \text{ seconds} - 13.14 \text{ seconds}$$
$$+ \left(1.2 \times 10^{-1} \text{ seconds}\right) = ?$$

Solve It

18. Express the answer to this calculation using the appropriate number of significant figures:

$$345.6 \text{ feet} \times \left(\frac{12 \text{ inches}}{1 \text{ foot}} \right) = ?$$

Solve It

19. Report the difference using the appropriate number of significant figures:

$$\left(3.7 \times 10^{-4} \text{ minutes}\right) - 0.009 \text{ minutes} = ?$$

Solve It

20. Express the answer to this multi-step calculation using the appropriate number of significant figures:

$$\frac{87.95 \text{ feet} \times 0.277 \text{ feet} + 5.02 \text{ feet} - 1.348 \text{ feet}}{10.0 \text{ feet}} = ?$$

Solve It

Answers to Questions on Noting Numbers Scientifically

The following are the answers to the practice problems in this chapter.

1 2×10^5. Move the decimal point immediately after the 2 to create a coefficient between 1 and 10. Because you're moving the decimal point five places to the left, multiply the coefficient, 2, by the power 10^5.

2 8.0736×10^4. Move the decimal point immediately after the 8 to create a coefficient between 1 and 10. You're moving the decimal point four places to the left, so multiply the coefficient, 8.0736, by the power 10^4.

3 2×10^{-5}. Move the decimal point immediately after the 2 to create a coefficient between 1 and 10. You're moving the decimal point five spaces to the right, so multiply the coefficient, 2, by the power 10^{-5}.

4 **690.3.** You need to understand scientific notation to change the number back to regular decimal form. Because 10^2 equals 100, multiply the coefficient, 6.903, by 100. This moves the decimal point two spaces to the right.

5 1.1×10^6. First, multiply the coefficients: $2.2 \times 5.0 = 11$. Then multiply the powers of ten by adding the exponents: $10^9 \times 10^{-4} = 10^{9+(-4)} = 10^5$. The raw calculation yields 11×10^5, which converts to the given answer when you express it in scientific notation.

6 3.0×10^{-7}. The ease of math with scientific notation shines through in this problem. Dividing the coefficients yields a coefficient quotient of $9.3/3.1 = 3.0$, and dividing the powers of ten (by subtracting their exponents) yields a quotient of $10^{-5}/10^2 = 10^{-5-2} = 10^{-7}$. Marrying the two quotients produces the given answer, already in scientific notation.

7 **1.8.** First, convert each number to scientific notation: 5.2×10^1 and 3.5×10^{-2}. Next, multiply the coefficients: $5.2 \times 3.5 = 18.2$. Then add the exponents on the powers of 10: $10^{1+(-2)} = 10^{-1}$. Finally, join the new coefficient with the new power: 18.2×10^{-1}. Expressed in scientific notation, this answer is $1.82 \times 10^0 = 1.82$. Looking back at the original numbers, you see that both factors have only two significant figures; therefore, you have to round your answer to match that number of sig figs, making it 1.8.

8 3.99×10^{-4}. First, convert each number to scientific notation: 8.09×10^{-3} and 2.03×10^1. Then divide the coefficients: $8.09/2.03 = 3.99$. Next, subtract the exponent on the denominator from the exponent of the numerator to get the new power of 10: $10^{-3-1} = 10^{-4}$. Join the new coefficient with the new power: 3.99×10^{-4}. Finally, express gratitude that the answer is already conveniently expressed in scientific notation.

9 545×10^{-6}. Because the numbers are each already expressed with identical powers of 10, you can simply add the coefficients: $398 + 147 = 545$. Then join the new coefficient with the original power of 10.

10 6.402×10^5. Because the numbers are each expressed with the same power of 10, you can simply subtract the coefficients: $7.685 - 1.283 = 6.402$. Then join the new coefficient with the original power of 10.

11 40.16×10^{-3} **(or an equivalent expression).** First, convert the numbers so they each use the same power of 10: 2.06×10^{-3} and 38.1×10^{-3}. Here, we use 10^{-3}, but you can use a different power as long as the power is the same for each number. Next, add the coefficients: $2.06 + 38.1 = 40.16$. Finally, join the new coefficient with the shared power of 10.

12 89.21×10^2 **(or an equivalent expression).** First, convert the numbers so each uses the same power of 10: 93.52×10^2 and 4.31×10^2. Here, we've picked 10^2, but any power is fine as long as the two numbers have the same power. Then subtract the coefficients: $93.52 - 4.31 = 89.21$. Finally, join the new coefficient with the shared power of 10.

13   **Reginald's measurement incurred the greater magnitude of error, and Dagmar's measurement incurred the greater percent error.** Reginald's scale reported with an error of 256 pounds – 237 pounds = 19 pounds, and Dagmar's scale reported with an error of 129 pounds – 117 pounds = 12 pounds. Comparing the *magnitudes* of error, you see that 19 pounds is greater than 12 pounds. However, Reginald's measurement had a percent error of (19 pounds/256 pounds)×100 = 7.4%, while Dagmar's measurement had a percent error of (12 pounds/129 pounds)×100 = 9.3%.

14   Jeweler A's official average measurement was 0.864 grams, and Jeweler B's official measurement was 0.856 grams. You determine these averages by adding up each jeweler's measurements and then dividing by the total number of measurements, in this case 3. Based on these averages, **Jeweler B's official measurement is more accurate** because it's closer to the actual value of 0.856 grams.

However, **Jeweler A's measurements were more precise** because the differences between A's measurements were much smaller than the differences between B's measurements. Despite the fact that Jeweler B's average measurement was closer to the actual value, the *range* of his measurements (that is, the difference between the largest and the smallest measurements) was 0.041 grams (0.875 g – 0.834 g = 0.041 g). The range of Jeweler A's measurements was 0.010 grams (0.869 g – 0.859 g = 0.010 g).

This example shows how low-precision measurements can yield highly accurate results through averaging of repeated measurements. In the case of Jeweler A, the error in the official measurement was 0.864 g – 0.856 g = 0.008 g. The corresponding percent error was (0.008 g/0.856 g)×100 = 0.9%. In the case of Jeweler B, the error in the official measurement was 0.856 g – 0.856 g = 0.000 g. Accordingly, the percent error was 0%.

15   The correct number of significant figures is as follows for each measurement: **a) 5, b) 3,** and **c) 4.**

16   **a) "893.7 $\pm$ 1 gram" is an improperly reported measurement because the reported value, 893.7, suggests that the measurement is certain to within a few tenths of a gram.** The reported error is known to be greater, at ±1 gram. The measurement should be reported as "894 ± 1 gram."

b) "342 $\pm$ 0.01 gram" is improperly reported because the reported value, 342, gives the impression that the measurement becomes uncertain at the level of grams. The reported error makes clear that uncertainty creeps into the measurement only at the level of hundredths of a gram. The measurement should be reported as "342.00 ± 0.01 gram."

17   **114.36 seconds.** The trick here is remembering to convert all measurements to the same power of 10 before comparing decimal places for significant figures. Doing so reveals that 1.2×10^{-1} seconds goes to the hundredths of a second, despite the fact that the measurement contains only two significant figures. The raw calculation yields 114.359 seconds, which rounds properly to the hundredths place (taking significant figures into account) as 114.36 seconds, or 1.1436×10^2 seconds in scientific notation.

18 4.147×10^3 **inches.** Here, you have to recall that defined quantities (1 foot is defined as 12 inches) have unlimited significant figures. So your calculation is limited only by the number of significant figures in the measurement 345.6 feet. When you multiply 345.6 feet by 12 inches per foot, the feet cancel, leaving units of inches:

$$\left(345.6 \ \cancel{ft}\right) \times \left(\frac{12 \text{ in.}}{1 \ \cancel{ft}}\right) = 4{,}147.2 \text{ in.}$$

The raw calculation yields 4,147.2 inches, which rounds properly to four significant figures as 4,147 inches, or 4.147×10^3 inches in scientific notation.

19   **–0.009 minutes.** Here, it helps here to convert all measurements to the same power of 10 so you can more easily compare decimal places in order to assign the proper number of significant figures. Doing so reveals that 3.7×10^{-4} minutes goes to the hundred-thousandths of a minute, and 0.009 minutes goes to the thousandths of a minute. The raw calculation yields –0.00863 minutes, which rounds properly to the thousandths place (taking significant figures into account) as –0.009 minutes, or -9×10^{-3} minutes in scientific notation.

20   **2.81 feet.** Following standard order of operations, you can do this problem in two main steps, first performing multiplication and division and then performing addition and subtraction.

Following the rules of significant-figure math, the first step yields 24.4 feet + 5.02 feet – 1.348 feet. Each product or quotient contains the same number of significant figures as the number in the calculation with the fewest number of significant figures.

After completing the first step, divide by 10.0 feet to finish the problem:

$$\frac{28.07 \text{ ft}}{10.0 \text{ ft}} = 2.807 \text{ ft} = 2.81 \text{ ft}$$

You write the answer with three sig figs because the measurement 10.0 feet contains three sig figs, which is the smallest available between the two numbers.

Chapter 2

Using and Converting Units

*H*ave you ever been asked for your height in centimeters, your weight in kilograms, or the speed limit in kilometers per hour? These measurements may seem a bit odd to those folks who are used to feet, pounds, and miles per hour, but the truth is that scientists sneer at feet, pounds, and miles. Because scientists around the globe constantly communicate numbers to each other, they prefer a highly systematic, standardized system. The *International System of Units,* abbreviated *SI* from the French term *Système International,* is the unit system of choice in the scientific community.

In this chapter, you find that the SI system offers a very logical and well-organized set of units. Scientists, despite what many of their hairstyles may imply, love logic and order, so SI is their system of choice.

As you work with SI units, try to develop a good sense for how big or small the various units are. That way, as you're doing problems, you'll have a sense for whether your answer is reasonable.

Familiarizing Yourself with Base Units and Metric System Prefixes

The first step in mastering the SI system is to figure out the base units. Much like the atom, the SI base units are building blocks for more-complicated units. In later sections of this chapter, you find out how more-complicated units are built from the SI base units. The five SI base units that you need to do chemistry problems (as well as their familiar, non-SI counterparts) are in Table 2-1.

Table 2-1			SI Base Units
Measurement	*SI Unit*	*Symbol*	*Non-SI Unit*
Amount of a substance	mole	mol	no non-SI unit
Length	meter	m	foot, inch, yard, mile
Mass	kilogram	kg	pound
Temperature	kelvin	K	degree Celsius, degree Fahrenheit
Time	second	s	minute, hour

Chemists routinely measure quantities that run the gamut from very small (the size of an atom, for example) to extremely large (such as the number of particles in one mole). Nobody, not even chemists, likes dealing with scientific notation (which we cover in Chapter 1) if they don't have to. For these reasons, chemists often use a metric system *prefix* (a word part that goes in front of the base unit to indicate a numerical value) in lieu of scientific notation. For example, the size of the nucleus of an atom is roughly 1 *nano*meter across, which is a nicer way of saying 1×10^{-9} meters across. The most useful of these prefixes are in Table 2-2.

Table 2-2		Metric System Prefixes	
Prefix	*Symbol*	*Meaning*	*Example*
kilo-	k	10^3	$1 \text{ km} = 10^3 \text{ m}$
deca-	D or da	10^1	$1 \text{ Dm} = 10^1 \text{ m}$
base unit	varies	1	1 m
deci-	d	10^{-1}	$1 \text{ dm} = 10^{-1} \text{ m}$
centi-	c	10^{-2}	$1 \text{ cm} = 10^{-2} \text{ m}$
milli-	m	10^{-3}	$1 \text{ mm} = 10^{-3} \text{ m}$
micro-	μ	10^{-6}	$1 \text{ μm} = 10^{-6} \text{ m}$
nano-	n	10^{-9}	$1 \text{ nm} = 10^{-9} \text{ m}$

Feel free to refer to Table 2-2 as you do your problems. You may want to earmark this page because after this chapter, we simply assume that you know how many meters are in 1 kilometer, how many grams are in 1 microgram, and so on.

Q. You measure a length to be 0.005 m. How can this be better expressed using a metric system prefix?

A. **5 mm.** 0.005 is 5×10^{-3} m, or 5 mm.

1. How many nanometers are in 1 cm?

Solve It

2. Your lab partner has measured the mass of your sample to be 2,500 g. How can you record this more nicely (without scientific notation) in your lab notebook using a metric system prefix?

Solve It

Building Derived Units from Base Units

Chemists aren't satisfied with measuring length, mass, temperature, and time alone. On the contrary, chemistry often deals in calculated quantities. These kinds of quantities are expressed with *derived units,* which are built from combinations of base units.

- ✔ **Area (for example, catalytic surface):** Area = Length × Width, and area has units of length squared (square meters, or m^2, for example).

- ✔ **Volume (of a reaction vessel, for example):** You calculate volume by using the familiar formula Volume = Length × Width × Height. Because length, width, and height are all length units, you end up with length × length × length, or a length cubed (for example, cubic meters, or m^3).

  The most common way of representing volume in chemistry is by using liters (L). You can treat the liter like you would any other metric base unit by adding prefixes to it, such as milli- or deci-.

- ✔ **Density (of an unidentified substance):** Density, arguably the most important derived unit to a chemist, is built by using the basic formula Density = Mass/Volume.

In the SI system, mass is measured in kilograms. The standard SI units for mass and length were chosen by the Scientific Powers That Be because many objects that you encounter in everyday life have masses between 1 and 100 kilograms and have dimensions on the order of 1 meter. Chemists, however, are most often concerned with very small masses and dimensions; in such cases, grams and centimeters are much more convenient. Therefore, the standard unit of density in chemistry is grams per cubic centimeter (g/cm^3) rather than kilograms per cubic meter.

The cubic centimeter is exactly equal to 1 milliliter, so densities are also often expressed in grams per milliliter (g/mL).

✔ **Pressure (of gaseous reactants, for example):** Pressure units are derived using the formula Pressure=Force/Area. The SI units for force and area are newtons (N) and square meters (m^2), so the SI unit of pressure, the pascal (Pa), can be expressed as N/m^2.

Q. A physicist measures the density of a substance to be 20 kg/m³. His chemist colleague, appalled with the excessively large units, decides to change the units of the measurement to the more familiar grams per cubic centimeter. What is the new expression of the density?

A. **0.02 g/cm³.** A kilogram contains 1,000 (10^3) grams, so 20 kg equals

20,000 g. Well, 100 cm = 1 m; therefore, (100 cm)³ = (1 m)³. In other words, there are 100^3 (or 10^6) cubic centimeters in 1 cubic meter. Doing the division gives you 0.02 g/cm³. You can write out the conversion as follows:

$$\left(\frac{20\text{ kg}}{1\text{ m}^3}\right)\left(\frac{10^3\text{ g}}{1\text{ kg}}\right)\left(\frac{1\text{ m}^3}{10^6\text{ cm}^3}\right)=0.02\,\frac{\text{g}}{\text{cm}^3}$$

3. The pascal, a unit of pressure, is equivalent to 1 newton per square meter. If the newton, a unit of force, is equal to a kilogram-meter per second squared, what is the pascal expressed entirely in basic units?

Solve It

4. A student measures the length, width, and height of a sample to be 10 mm, 15 mm, and 5 mm, respectively. If the sample has a mass of 0.9 Dg, what is the sample's density in grams per milliliter?

Solve It

Converting between Units: The Conversion Factor

So what happens when chemist Reginald Q. Geekmajor neglects his SI units and measures the boiling point of his sample to be 101 degrees Fahrenheit or measures the volume of his beaker to be 2 cups? Although Dr. Geekmajor should surely have known better, he can still save himself the embarrassment of reporting such dirty, unscientific numbers to his colleagues: He can use conversion factors.

A *conversion factor* simply uses your knowledge of the relationships between units to convert from one unit to another. For example, if you know that there are 2.54 centimeters in every inch (or 2.2 pounds in every kilogram or 101.3 kilopascals in every atmosphere), then converting between those units becomes simple algebra. Peruse Table 2-3 for some useful conversion factors. And remember: If you know the relationship between any two units, you can build your own conversion factor to move between those units.

Table 2-3	Conversion Factors	
Unit	*Equivalent To*	*Conversion Factors*
Length		
1 meter	3.3 feet	$\frac{3.3\,\text{ft}}{1\,\text{m}}$ or $\frac{1\,\text{m}}{3.3\,\text{ft}}$
1 foot	12 inches	$\frac{1\,\text{ft}}{12\,\text{in.}}$ or $\frac{12\,\text{in.}}{1\,\text{ft}}$
1 inch	2.54 centimeters	$\frac{1\,\text{in.}}{2.54\,\text{cm}}$ or $\frac{2.54\,\text{cm}}{1\,\text{in.}}$
Volume		
1 gallon	16 cups	$\frac{16\,\text{c}}{1\,\text{gal}}$ or $\frac{1\,\text{gal}}{16\,\text{c}}$
1 cup	237 milliliters	$\frac{1\,\text{c}}{237\,\text{mL}}$ or $\frac{237\,\text{mL}}{1\,\text{c}}$
1 milliliter	1 cubic centimeter	$\frac{1\,\text{mL}}{1\,\text{cm}^3}$ or $\frac{1\,\text{cm}^3}{1\,\text{mL}}$
Mass		
1 kilogram	2.2 pounds	$\frac{1\,\text{kg}}{2.2\,\text{lb}}$ or $\frac{2.2\,\text{lb}}{1\,\text{kg}}$
Time		
1 hour	3,600 seconds	$\frac{1\,\text{hr}}{3,600\,\text{s}}$ or $\frac{3,600\,\text{s}}{1\,\text{hr}}$
Pressure		
1 atmosphere	101.3 kilopascals	$\frac{1\,\text{atm}}{101.3\,\text{kPa}}$ or $\frac{101.3\,\text{kPa}}{1\,\text{atm}}$
1 atmosphere	760 millimeters of mercury (mm Hg*)	$\frac{1\,\text{atm}}{760\,\text{mm Hg}}$ or $\frac{760\,\text{mm Hg}}{1\,\text{atm}}$

* One of the more peculiar units you'll encounter in your study of chemistry is mm Hg, or millimeters of mercury, a unit of pressure. Unlike SI units, mm Hg doesn't fit neatly into the base-10 metric system, but it reflects the way in which certain devices like blood pressure cuffs and barometers use mercury to measure pressure.

As with many things in life, chemistry isn't always as easy as it seems. Chemistry teachers are sneaky: They often give you quantities in non-SI units and expect you to use one or more conversion factors to change them to SI units — all this before you even attempt the "hard part" of the problem! We're at least marginally less sneaky than your typical chemistry teacher, but we hope to prepare you for such deception, so expect to use conversion factors throughout the rest of this book!

The following example shows how to use a basic conversion factor to fix non-SI units.

Q. Dr. Geekmajor absentmindedly measures the mass of a sample to be 0.75 lb and records his measurement in his lab notebook. His astute lab assistant, who wants to save the doctor some embarrassment, knows that there are 2.2 lb in every kilogram. The assistant quickly converts the doctor's measurement to SI units. What does she get?

A. **0.34 kg.**

$$\left(0.75 \, \cancel{lb}\right)\left(\frac{1 \, kg}{2.2 \, \cancel{lb}}\right) = 0.34 \, kg$$

Notice that something very convenient happens because of the way this calculation is set up. In algebra, whenever you find the same quantity in a numerator and in a denominator, you can cancel it out. Canceling out the pounds (lb) is a lovely bit of algebra because you don't want those units around, anyway. The whole point of the conversion factor is to get rid of an undesirable unit, transforming it into a desirable one — without breaking any rules. Always let the units be your guide when solving a problem. Ensure the right ones cancel out, and if they don't, go back and flip your conversion factor.

If you're a chemistry student, you're probably pretty familiar with the basic rules of algebra. So you know that you can't simply multiply one number by another and pretend that nothing happened — you altered the original quantity when you multiplied, didn't you? Thankfully, the answer is *no*. You're simply referring to the amount using different units. A meter stick is always going to be the same length, whether you say it's 1 m long or 100 cm long. The physical length of the stick hasn't changed, despite your using a different unit to describe it.

Recall another algebra rule: You can multiply any quantity by 1, and you'll always get back the original quantity. Now look closely at the conversion factors in the example: 2.2 lb and 1 kg are exactly the same thing! Multiplying by 2.2 lb/1 kg or by 1 kg/2.2 lb is really no different from multiplying by 1.

Q. A chemistry student, daydreaming during lab, suddenly looks down to find that he's measured the volume of his sample to be 1.5 cubic *inches*. What does he get when he converts this quantity to cubic centimeters?

A. **25 cm³.**

$$\left(1.5 \, in^3\right)\left(\frac{2.54 \, cm}{1 \, in}\right)^3 = \left(1.5 \, in^3\right)\left(\frac{16.39 \, cm^3}{1 \, in^3}\right) = 25 \, cm^3$$

Rookie chemists often mistakenly assume that if there are 2.54 centimeters in every inch, then there are 2.54 cubic centimeters in every cubic inch. No! Although this assumption seems logical at first glance, it leads to catastrophically wrong answers. Remember that cubic units are units of volume and that the formula for volume is Length×Width×Height. Imagine 1 cubic inch as a cube with 1-inch sides. The cube's volume is 1 in.×1 in.×1 in. = 1 in.^{3}.

Now consider the dimensions of the cube in centimeters: 2.54 cm×2.54 cm×2.54 cm. Calculate the volume using these measurements, and you get 2.54 cm×2.54 cm×2.54 cm = 16.39 cm^3. This volume is much greater than 2.54 cm^3! To convert units of area or volume using length measurements, square or cube everything in your conversion factor, not just the units, and everything works out just fine.

5. A sprinter running the 100.0 m dash runs how many feet?

Solve It

6. At the top of Mount Everest, the air pressure is approximately 0.330 atmospheres, or one-third of the air pressure at sea level. A barometer placed at the peak would read how many millimeters of mercury?

Solve It

7. A *league* is an obsolete unit of distance used by archaic (or nostalgic) sailors. A league is equivalent to 5.6 km. If the characters in Jules Verne's novel *20,000 Leagues Under the Sea* travel to a depth of 20,000 leagues, how many kilometers under the surface of the water are they? If the radius of the Earth is 6,378 km, is this a reasonable depth? Why or why not?

Solve It

8. The slab of butter that Paul Bunyan slathered on his morning pancakes is 2.0 ft wide, 2.0 ft long, and 1.0 ft thick. How many cubic meters of butter does Paul consume each morning?

Solve It

Letting the Units Guide You

You don't need to know all possible unit conversions (between meters and inches, for example). Instead of memorizing or looking up conversion factors between all types of units, you can memorize just a handful of conversion factors and use them one after another, letting the units guide you each step of the way.

Say you want to know the number of seconds in a standard calendar year (clearly a very large number, so don't forget about scientific notation, as we explain in Chapter 1). Very few people have this conversion memorized — or will admit to it — but everyone knows that there are 60 seconds in a minute, 60 minutes in an hour, 24 hours in a day, and 365 days in a standard calendar year. So use what you know to get what you want!

$$(1 \text{ yr}) \left(\frac{365 \text{ days}}{1 \text{ yr}} \right) \left(\frac{24 \text{ hr}}{1 \text{ day}} \right) \left(\frac{60 \text{ min}}{1 \text{ hr}} \right) \left(\frac{60 \text{ s}}{1 \text{ min}} \right) = 3.15 \times 10^7 \text{ s}$$

You can use as many conversion factors as you need as long as you keep track of your units in each step. The easiest way to do this is to cancel as you go, using the remaining unit as a guide for the next conversion factor. For example, examine the first two factors of the years-to-seconds conversion. The *years* on the top and bottom cancel, leaving you with *days*. Because *days* remains on top, the next conversion factor needs to have *days* on the bottom and *hours* on the top. Canceling *days* then leaves you with *hours*, so your next conversion factor must have *hours* on the bottom and *minutes* on the top. Just repeat this process until you arrive at the units you want. Then do all the multiplying and dividing of the numbers, and rest assured that the resulting calculation is the right one for the final units.

$$(1 \text{ yr}) \left(\frac{365 \text{ days}}{1 \text{ yr}} \right) \left(\frac{24 \text{ hr}}{1 \text{ day}} \right) \left(\frac{60 \text{ min}}{1 \text{ hr}} \right) \left(\frac{60 \text{ s}}{1 \text{ min}} \right) = 3.15 \times 10^7 \text{ s}$$

Q. A chemistry student measures a length of 423 mm, yet the lab she's working on requires that it be in kilometers. What is the length in kilometers?

A. **4.23×10^{-4} km.** You can go about solving this problem in two ways. We first show you the slightly longer way involving two conversions, and then we shorten it to a nice, simple one-step problem.

This conversion requires you to move across the metric-system prefixes you find in Table 2-2. When you're working on a conversion that passes through a base unit, it may be helpful to treat the process as two steps, converting to and from the base unit. In this case, you can convert from millimeters to meters and then from meters to kilometers:

$$(423 \text{ mm}) \left(\frac{1 \text{ m}}{1,000 \text{ mm}} \right) \left(\frac{1 \text{ km}}{1,000 \text{ m}} \right) = 4.23 \times 10^{-4} \text{ km}$$

You can see how *millimeters* cancels out, and you're left with *meters*. Then *meters* cancels out, and you're left with your desired unit, *kilometers*.

The second way you can approach this problem is to treat the conversion from milli- to kilo- as one big step:

$$(423 \text{ mm}) \left(\frac{1 \text{ km}}{10^6 \text{ mm}} \right) = 4.23 \times 10^{-4} \text{ km}$$

Notice the answer doesn't change; the only difference is the number of steps required to convert the units. Based on Table 2-2 and the first approach we showed you, you can see that the total conversion from millimeters to kilometers requires 10^6 mm to 1 km. You're simply combining the two denominators in the two-step conversion (1,000 mm and 1,000 m) into one. Rewriting each 1,000 as 10^3 may help you see how the denominators combine to become 10^6.

---

9. How many meters are in 15 ft?

Solve It

10. If Steve weighs 175 lb, what's his weight in grams?

Solve It

11. How many liters are in 1 gal of water?

Solve It

12. If the dimensions of a solid sample are 3 in. x 6 in. x 1 ft, what's the volume of that sample in cubic centimeters? Give your answer in scientific notation or use a metric prefix.

Solve It

13. If there are 5.65 kg per every half liter of a particular substance, is that substance liquid mercury (density 13.5 g/cm^3), lead (density 11.3 g/cm^3), or tin (density 7.3 g/cm^3)?

Solve It

Answers to Questions on Using and Converting Units

Following are the answers to the practice problems presented in this chapter.

1 1×10^7 **nm.** Both 10^2 centimeters and 10^9 nanometers equal 1 meter. Set the two measurements equal to one another (10^2 cm $= 10^9$ nm) and solve for centimeters by dividing. This conversion tells you that 1 cm $= 10^9/10^2$ nm, or 1×10^7 nm.

2 **2.5 kg.** Because 1,000 g are in 1 kg, simply divide 2,500 by 1,000 to get 2.5.

3 $1 \, \text{Pa} = \dfrac{1 \, \text{kg}}{\text{m} \cdot \text{s}^2}$. First, write out the equivalents of pascals and newtons as the problem explains:

$$1 \, \text{Pa} = \frac{1 \, \text{N}}{\text{m}^2} \quad \text{and} \quad 1 \, \text{N} = \frac{1 \, \text{kg} \cdot \text{m}}{\text{s}^2}$$

Now substitute *newtons* (expressed in fundamental units) into the equation for the *pascal:*

$$1 \, \text{Pa} = \frac{\dfrac{1 \, \text{kg} \cdot \text{m}}{\text{s}^2}}{\text{m}^2}$$

Simplify this equation to $1 \, \text{Pa} = \dfrac{1 \, \text{kg} \cdot \text{m}}{\text{m}^2 \cdot \text{s}^2}$ and cancel out the *meter,* which appears in both the top and the bottom, leaving $1 \, \text{Pa} = 1 \dfrac{\text{kg}}{\text{m} \cdot \text{s}^2}$.

4 **12 g/mL.** Because a milliliter is equivalent to a cubic centimeter, the first thing to do is to convert all the length measurements to centimeters: 1 cm, 1.5 cm, and 0.5 cm. Then multiply the converted lengths to get the volume: (1 cm)(1.5 cm)(0.5 cm) $= 0.75$ cm^3, or 0.75 mL. The mass should be expressed in grams rather than decagrams; there are 10 grams in 1 decagram, so 0.9 Dg $= 9$ g. Using the formula $D = m/V$, you calculate a density of 9 grams per 0.75 milliliter, or 12 g/mL.

5 **330 ft.** Set up the conversion factor as follows:

$$(100.0 \, \text{m}) \left(\frac{3.3 \, \text{ft}}{1 \, \text{m}} \right) = 330 \, \text{ft}$$

6 **251 mm Hg.**

$$(0.330 \, \text{atm}) \left(\frac{760 \, \text{mm Hg}}{1 \, \text{atm}} \right) = 251 \, \text{mm Hg}$$

7 1.12×10^5 **km.**

$$(20{,}000 \, \text{leagues}) \left(\frac{5.6 \, \text{km}}{1 \, \text{league}} \right) = 1.12 \times 10^5 \, \text{km}$$

The radius of the Earth is only 6,378 km, and 20,000 leagues is 17.5 times that radius! So the ship would've burrowed through the Earth and been halfway to the orbit of Mars if it had truly sunk to such a depth. Jules Verne's title refers to the distance the submarine travels through the sea, not its depth.

8 **0.11 m³.** The volume of the butter in feet is $2.0\ \text{ft} \times 2.0\ \text{ft} \times 1.0\ \text{ft}$, or $4\ \text{ft}^3$.

$$\left(4.0\ \text{ft}^3\right)\left(\frac{1\ \text{m}}{3.3\ \text{ft}}\right)^3 = \left(4.0\ \text{ft}^3\right)\left(\frac{1\ \text{m}^3}{35.937\ \text{ft}^3}\right) = 0.11\ \text{m}^3$$

9 **4.6 m.** You have to convert all the way from feet to meters. Looking at the conversion factors in Table 2-3, you should see that you can convert feet into inches and then inches into centimeters. Then you can easily convert centimeters into meters.

$$(15\ \text{ft})\left(\frac{12\ \text{in.}}{1\ \text{ft}}\right)\left(\frac{2.54\ \text{cm}}{1\ \text{in.}}\right)\left(\frac{1\ \text{m}}{100\ \text{cm}}\right) = 4.6\ \text{m}$$

10 **7.95 × 10⁴ g.** There's no direct pound-to-gram conversion factor in Table 2-3, so you must determine the correct path to take. In this case, you can convert from pounds to kilograms and then from kilograms to grams:

$$(175\ \text{lb})\left(\frac{1\ \text{kg}}{2.2\ \text{lb}}\right)\left(\frac{1,000\ \text{g}}{1\ \text{kg}}\right) = 7.95 \times 10^4\ \text{g}$$

11 **3.8 L.** You must determine the correct pathway to get from gallons to liters using the conversions provided in Table 2-3. To do so, convert from gallons to cups, then to milliliters, and finally to liters:

$$(1\ \text{gal})\left(\frac{16\ \text{c}}{1\ \text{gal}}\right)\left(\frac{237\ \text{mL}}{1\ \text{c}}\right)\left(\frac{1\ \text{L}}{1,000\ \text{mL}}\right) = 3.8\ \text{L}$$

12 **3.54 × 10³ cm³.** First convert all the inch and foot measurements to centimeters:

$$(3\ \text{in})\left(\frac{2.54\ \text{cm}}{1\ \text{in}}\right) = 7.62\ \text{cm}$$

$$(6\ \text{in})\left(\frac{2.54\ \text{cm}}{1\ \text{in}}\right) = 15.24\ \text{cm}$$

$$(1\ \text{ft})\left(\frac{12\ \text{in}}{1\ \text{ft}}\right)\left(\frac{2.54\ \text{cm}}{1\ \text{in}}\right) = 30.48\ \text{cm}$$

The volume is therefore $7.62\ \text{cm} \times 15.24\ \text{cm} \times 30.48\ \text{cm}$, or $3.54 \times 10^3\ \text{cm}^3$.

13 **The substance is lead.** To set up this problem, be sure to begin with the correct initial amounts. The problem tells you there are 5.65 kg for every half liter of substance. This translates into 5.65 kg/0.5 L. After you've established the initial value, use conversion factors to find the density:

$$\left(\frac{5.65\ \text{kg}}{0.5\ \text{L}}\right)\left(\frac{1,000\ \text{g}}{1\ \text{kg}}\right)\left(\frac{1\ \text{L}}{1,000\ \text{mL}}\right)\left(\frac{1\ \text{mL}}{1\ \text{cm}^3}\right) = 11.3\ \frac{\text{g}}{\text{cm}^3}$$

This answer is exactly the density of lead.

Chapter 3

Breaking Down Atoms

. . . .

In This Chapter

▶ Peeking inside the atom: Protons, electrons, and neutrons

▶ Deciphering atomic numbers and mass numbers

▶ Understanding isotopes and calculating atomic masses

. . . .

"**B**ig stuff is built from smaller pieces of stuff. If you keep breaking stuff down into smaller and smaller pieces, eventually you'll reach the smallest possible bit of stuff. Let's call that bit an *atom*." This is how the Greek philosopher Democritus might have explained his budding concept of "atomism" to a buddy over a flask of Cretan wine. Like wine, the idea had legs.

For hundreds of years, scientists have operated under the idea that all matter is made up of smaller building blocks called *atoms*. So small, in fact, that until the invention of the electron microscope in 1931, the only way to find out anything about these tiny, mysterious particles was to design a very, very clever experiment. Chemists couldn't exactly corner a single atom in a back alley somewhere and study it alone — they had to study the properties of whole gangs of atoms and try to guess what individual ones might be like. Through remarkable ingenuity and incredible luck, chemists now understand a great deal about the atom. After reading this chapter, so will you.

The Atom: Protons, Electrons, and Neutrons

Picture an atom as a microscopic LEGO. Atoms come in a variety of shapes and sizes, and you can build larger structures out of them. Like a LEGO, an atom is extremely hard to break. In fact, so much energy is stored inside atoms that breaking them in half results in a nuclear explosion.

Even though an atom is made of smaller pieces called *subatomic particles,* the atom is still considered the smallest possible unit of an element, because after you break an atom of an element into subatomic particles, the pieces lose the unique properties of that element.

Virtually all substances are made of atoms. The universe seems to use about 120 unique atomic LEGO blocks to build neat things like galaxies and people and whatnot. All atoms are made of the same three subatomic particles: the proton, the electron, and the neutron. Different types of atoms (in other words, different *elements*) have different combinations of these particles, which gives each element unique properties. For example:

- ✔ **Atoms of different elements have different masses.** Atomic masses are measured in multiples of the mass of a single proton, called *atomic mass units* (amu), which are equivalent to 1.66×10^{-27} kg. We discuss atomic mass in more detail in the later section "Accounting for Isotopes Using Atomic Masses."

- ✔ **Two of the subatomic particles have a charge: The proton is positive, and the electron is negative.** Atomic charges are measured in multiples of the charge of a single proton.

The must-know information about the three subatomic particles is in Table 3-1. Notice that protons and electrons have equal and opposite charges and that neutrons are neutral. Atoms always have an equal number of protons and electrons, so the overall charge of an atom is neutral (that is to say, zero).

Many atoms actually prefer to gain or lose electrons, which gives them a nonzero charge; in other words, the number of negative charges is no longer balanced by the number of positive charges. Charged atoms are called *ions* (see Chapter 5). For now, we'll deal only with atoms that have equal numbers of protons and electrons.

Table 3-1	Subatomic Particles	
Particle	*Mass*	*Charge*
Proton	1 amu	+1
Electron	$\frac{1}{1,836}$ amu	−1
Neutron	1 amu	0

Now look at Table 3-1 with an eye to mass. A proton and a neutron have approximately the same mass, which is about 2,000 times the mass of an electron. This means that most of an atom's mass comes from protons and neutrons. Although electrons contribute a lot of negative charge, they contribute very little mass to an atom.

An atom's nucleus, the center of the atom, contains protons and neutrons; therefore, protons and neutrons are sometimes called *nuclear particles*. Electrons move around the nucleus in a cloud of many different energy levels.

Q. If a gold atom contains 197 nuclear particles, 79 of which are protons, how many neutrons and how many electrons does the gold atom have?

A. **118 neutrons and 79 electrons.** The *nucleus* contains all the protons and neutrons in an atom, so if 79 of the 197 particles in a gold nucleus are protons, the remaining 118 particles must be neutrons. All atoms are electrically neutral, so there must be a total of 79 electrons (in other words, 79 negative charges) to balance out the 79 positive charges of the protons. This type of logic leads to a general formula that you can use to calculate proton or neutron counts. This formula is $M = P + N$, where M is the atomic mass, P is the number of protons, and N is the number of neutrons.

1. If an atom has 71 protons, 71 electrons, and 104 neutrons, how many particles reside in the nucleus, and how many are outside of the nucleus?

Solve It

2. If an atom's nucleus has a mass of 31 amu and contains 15 protons, how many neutrons and electrons does the atom have?

Solve It

Deciphering Chemical Symbols: Atomic and Mass Numbers

Two very important numbers, the atomic number and the mass number, tell you much of what you need to know about an atom. Chemists tend to memorize these numbers like baseball fans memorize batting averages, but clever chemistry students like you need not resort to memorization. You have the ever-important periodic table of the elements at your disposal. We discuss the logical structure and organization of the periodic table in detail in Chapter 4, so for now, we simply explain what the atomic and mass numbers mean without going into great detail about their consequences.

Atomic numbers are like name tags: They identify an element as carbon, nitrogen, beryllium, and so on by telling you the number of protons in the nucleus of that element. Atoms are known by the numbers of their protons. Adding a proton or removing one from the nucleus of an atom changes the elemental identity of an atom.

In the periodic table, you find the atomic number above the one- or two-letter abbreviation for an element. The abbreviation is the element's chemical symbol. Notice that the elements of the periodic table are lined up in order of atomic number, as if they've responded to some sort of roll call. Atomic number increases by 1 each time you move to the right in the periodic table; when a row ends, the sequence of increasing atomic numbers begins again at the left side of the next row down. You can check out the periodic table for yourself in Chapter 4.

The second identifying number of an atom is its mass number. The mass number reports the mass of the atom's nucleus in atomic mass units (amu). Because protons and neutrons have a mass of 1 amu each (as you find out earlier in this chapter), the mass number equals the sum of the numbers of protons and neutrons:

$$\text{Mass Number} = \text{Protons} + \text{Neutrons}$$

Why, you may wonder, don't we care about the mass of the electrons? Is some sort of insidious subatomic particle-ism afoot? No. An electron has only 1/1,836 of the mass of a proton or neutron, so to make mass numbers nice and even, chemists have decided to conveniently forget that electrons have mass. Although this assumption is not, well, true, the contributions of electrons to the mass of an atom are so small that the assumption is usually harmless. Electron mass is accounted for at the upper levels of chemistry, however, so don't worry.

To specify the atomic and mass numbers of an element, chemists typically write the symbol of the element in the form $^{A}_{Z}X$, where Z is the atomic number, A is the mass number, and X is the chemical symbol for that element. This form is often called *isotope notation.*

Q. What are the name, atomic number, mass number, number of protons, number of electrons, and number of neutrons of each of the following four elements: $^{35}_{17}Cl$, $^{37}_{17}Cl$, $^{190}_{76}Os$, and $^{39}_{19}K$?

A. The answers to questions like these, favorites of chemistry teachers, are best organized in a table. First, look up the symbols Cl, Os, and K in the periodic table in Chapter 4 and find the names of these elements. Enter what you find in the first column. To fill in the second and third columns (Atomic Number and Mass Number), read the atomic number and mass number from the lower left and upper left of the chemical symbols given in the question. The atomic number equals the number of protons; the number of electrons is the same as the number of protons, because elements have zero overall charge. So fill in the proton and electron columns with the same numbers you entered in column two. Last, subtract the atomic number from the mass number to get the number of neutrons, and enter that value in column six. Voilà! The entire private life of each of these atoms is now laid before you. Your answer should look like the following table.

Element Name	Atomic Number	Mass Number	Number of Protons	Number of Electrons	Number of Neutrons
Chlorine	17	35	17	17	18
Chlorine	17	37	17	17	20
Osmium	76	190	76	76	114
Potassium	19	39	19	19	20

3. Write the proper chemical symbol for an atom of bismuth with a mass of 209 amu.

Solve It

---

4. Fill in the following chart for $^{1}_{1}H$, $^{52}_{24}Cr$, $^{192}_{77}Ir$, and $^{96}_{42}Mo$.

Element Name	Atomic Number	Mass Number	Number of Protons	Number of Electrons	Number of Neutrons

---

5. Write the $^{A}_{Z}X$ form of the two elements in the following table:

Element Name	Atomic Number	Mass Number	Number of Protons	Number of Electrons	Number of Neutrons
Tungsten	74	184	74	74	110
Lead	82	207	82	82	125

---

6. Use the periodic table and your knowledge of atomic numbers and mass numbers to fill in the missing pieces in the following table.

Name	Atomic Number	Mass Number	Number of Protons	Number of Electrons	Number of Neutrons
Silver		108			
	16				16
		64	29		
				18	22

Accounting for Isotopes Using Atomic Masses

It's Saturday night. The air is charged with possibility. Living in the moment, you peruse your personal copy of the periodic table. What's this? You notice the numbers that appear below the atomic symbols appear to be related to the elements' mass numbers — but they're not nice whole numbers. What could it all mean?

In the preceding section, we explain that the mass number of an atom equals the sum of the numbers of protons and neutrons in the atom's nucleus. So how can you have a fractional number of protons or neutrons? There's no such thing as half a proton or 0.25 neutrons. Your chemist's sense of meticulousness has been offended, and you demand answers.

As it turns out, most elements come in several different varieties, called *isotopes*. Isotopes are atoms of the same element that have different mass numbers; the differences in mass number arise from different numbers of neutrons. The messy looking numbers with all those decimal places are atomic masses. An *atomic mass* is a weighted average of the masses of all the naturally occurring isotopes of an element. Chemists have measured the percentage of each element that exists in different isotopic forms. In the weighted average of the atomic mass, the mass of each isotope contributes in proportion to how often that isotope occurs in nature. More-common isotopes contribute more to the atomic mass.

Consider the element carbon, for example. Carbon occurs naturally in three isotopes:

- **Carbon-12** ($^{12}_6C$, or carbon with six protons and six neutrons) is boring old run-of-the-mill carbon, accounting for 99 percent of all the carbon out there.

- **Carbon-13** ($^{13}_6C$, or carbon with six protons and seven neutrons) is a slightly rarer (though still dull) isotope, accounting for most of the remaining 1 percent of carbon atoms. Taking on an extra neutron makes carbon-13 slightly heavier than carbon-12 but does little else to change its properties.

  However, even this minor change has some useful consequences. Scientists compare the ratio of carbon-12 to carbon-13 within meteorites to help determine their origins.

- **Carbon-14** ($^{14}_6C$, or carbon with six protons and eight neutrons) shows its interesting little face in only one out of every trillion or so carbon atoms. So if you're thinking you won't be working with large samples of carbon-14 in your chemistry lab, you're right!

These three isotopes are why you see carbon's atomic mass on the periodic table written as 12.01. If you do a quick bit of deductive reasoning, you can probably determine that carbon-12 is far and away the most common of the three isotopes due to the average atomic mass being closest to 12.

Precise measurements of the amounts of different isotopes can be important. You need to know the exact measurements if you're asked to figure out an element's atomic mass. To calculate an atomic mass, you need to know the masses of the isotopes and the percentage of the element that occurs as each isotope (this is called the *relative abundance*). To calculate an average atomic mass, make a list of each isotope along with its mass and its percent relative abundance. Multiply the mass of each isotope by its relative abundance. Add the products. The resulting sum is the atomic mass.

Certain elements, such as chlorine, occur in several very common isotopes, so their average atomic mass isn't close to a whole number. Other elements, such as carbon, occur in one very common isotope and several very rare ones, resulting in an average atomic mass that's very close to the whole-number mass of the most common isotope.

Q. Chlorine occurs in two common isotopes. It appears as $^{35}_{17}Cl$ 75.8% of the time and as $^{37}_{17}Cl$ 24.2% of the time. What is its average atomic mass?

A. **35.5 amu.** First, multiply each atomic mass by its relative abundance, using the decimal form of your percentage (75.8% = 0.758):

$$(35 \text{ amu})(0.758) = 26.53 \text{ amu}$$
$$(37 \text{ amu})(0.242) = 8.95 \text{ amu}$$

Then add the two results together to get the average atomic mass:

$$26.53 \text{ amu} + 8.95 \text{ amu} = 35.5 \text{ amu}$$

Compare your answer to the value on your periodic table. If you've done the calculation correctly, the two values should match or at least be very, very similar.

---

7. Magnesium occurs in three fairly common isotopes, $^{24}_{12}Mg$, $^{25}_{12}Mg$, and $^{26}_{12}Mg$, which have percent abundances of 78.9%, 10.0%, and 11.1%, respectively. Calculate the average atomic mass of magnesium.

Solve It

Answers to Questions on Atoms

Now that you've organized the facts on atoms into clearly labeled compartments in your brain, see how accessible that information is. Check your answers to the practice problems presented in this chapter.

1 **175 inside, 71 outside.** The nucleus of an atom consists of protons and neutrons, so this atom (lutetium) has 175 particles in its nucleus (71 protons + 104 neutrons). Electrons are the only subatomic particles that aren't included in the nucleus, so lutetium has 71 particles outside of its nucleus.

2 **16 neutrons, 15 electrons.** A nuclear mass of 31 amu means that the nucleus has 31 particles. Because 15 of them are protons, that leaves 16 amu for the neutrons. The numbers of protons and electrons are equal in a neutral atom (in this case, we're talking about phosphorus), so the atom has 15 electrons.

3 $^{209}_{83}$**Bi.** Find bismuth on the periodic table to get its chemical abbreviation and its atomic number. Because you already have its mass number (209), all you need to do is write all this information in $^{A}_{Z}X$ form.

4 To fill out the chart, take the information presented in the isotope notation and break it into the individual pieces the chart asks for. Looking at hydrogen, you see that it has a mass number of 1 and an atomic number of 1. This means only one proton is present and no neutrons are present. The number of electrons is 1 because in a neutral atom, the number of electrons is always equal to the number of protons.

To solve chromium's line, notice that chromium has a mass number of 52 and an atomic number of 24. This tells you that the number of protons is 24, because the atomic number and the number of protons are the same. To determine the number of neutrons, you simply subtract the atomic number from the mass number: $52 - 24 = 28$ neutrons. Finally, the number of protons is equal to the number of electrons. Continue this process to fill in the info for iridium and molybdenum.

Element Name	Atomic Number	Mass Number	Number of Protons	Number of Electrons	Number of Neutrons
Hydrogen	1	1	1	1	0
Chromium	24	52	24	24	28
Iridium	77	192	77	77	115
Molybdenum	42	96	42	42	54

5 $^{184}_{74}$**W**, $^{207}_{82}$**Pb.** Notice that you don't need most of the information in the table; all you really need to look up is the chemical symbol of each element.

6 When you're given the atomic number, the number of protons, or the number of electrons, you automatically know the other two numbers because they're all equal. Each element in the periodic table is listed with its atomic number, so by locating the element, you can simply read off the atomic number and therefore know the number of protons and electrons. To calculate the atomic mass or the number of neutrons, you must be given one or the other. Calculate atomic mass by adding the number of protons to the number of neutrons. Alternatively, calculate the number of neutrons by subtracting the number of protons from the atomic mass.

Element Name	Atomic Number	Mass Number	Number of Protons	Number of Electrons	Number of Neutrons
Silver	47	108	47	47	61
Sulfur	16	32	16	16	16
Copper	29	64	29	29	35
Argon	18	40	18	18	22

7 **24.3 amu.** First, multiply the three mass numbers by their relative abundances in decimal form. Then add the resulting products to get the average atomic mass.

$$(24 \text{ amu})(0.789) = 18.94 \text{ amu}$$
$$(25 \text{ amu})(0.100) = 2.50 \text{ amu}$$
$$(26 \text{ amu})(0.111) = 2.89 \text{ amu}$$
$$18.94 \text{ amu} + 2.50 \text{ amu} + 2.89 \text{ amu} = 24.3 \text{ amu}$$

Chapter 4

Surveying the Periodic Table of the Elements

. . . .

In This Chapter

▶ Moving up, down, left, and right on the periodic table

▶ Predicting elements' properties from periods and groups

▶ Grasping the value of valence electrons

▶ Taking stock of electron configurations

▶ Equating an electron's energy to light

. . . .

There it hangs, looming ominously over the chemistry classroom, a formidable wall built of bricks with names like "C," "Ag," and "Tc." Behold the *periodic table of the elements!* But don't be fooled by its stern appearance or intimidated by its teeming details. The table is your friend, your guide, your key to making sense of chemistry. To begin to make friends with the table, concentrate on its trends. Start simply: Notice that the table has rows and columns. Keep your eye on the columns and rows, and soon you'll be making sense of things like electron affinity and atomic radii. Really.

Organizing the Periodic Table into Periods and Groups

Take a look at your new friend, the periodic table, in Figure 4-1. Notice the horizontal rows and the vertical columns of elements:

> ✔ **Periods:** The rows are called *periods*. As you move across any period, you pass over a series of elements whose properties change in a predictable way.
>
> ✔ **Groups:** The columns are called *groups* or *families*. Notice the group label atop each column. The elements within any group have very similar properties. The properties of the elements emerge mostly from their different numbers of protons and electrons (see Chapter 3 for a refresher) and from the arrangement of their electrons.

We explain how to predict the properties of elements from looking at the periodic table in the next section. For now, we just want to describe how the elements are arranged in periods and groups.

PERIODIC TABLE OF THE ELEMENTS

1 IA	2 IIA	3 IIIB	4 IVB	5 VB	6 VIB	7 VIIB	8 VIIIB	9 VIIIB	10 VIIIB	11 IB	12 IIB	13 IIIA	14 IVA	15 VA	16 VIA	17 VIIA	18 VIIIA
1 H Hydrogen 1.00797																	2 He Helium 4.0026
3 Li Lithium 6.939	4 Be Beryllium 9.0122											5 B Boron 10.811	6 C Carbon 12.01115	7 N Nitrogen 14.0067	8 O Oxygen 15.9994	9 F Fluorine 18.9984	10 Ne Neon 20.183
11 Na Sodium 22.9898	12 Mg Magnesium 24.312											13 Al Aluminum 26.9815	14 Si Silicon 28.086	15 P Phosphorus 30.9738	16 S Sulfur 32.064	17 Cl Chlorine 35.453	18 Ar Argon 39.948
19 K Potassium 39.102	20 Ca Calcium 40.08	21 Sc Scandium 44.956	22 Ti Titanium 47.90	23 V Vanadium 50.942	24 Cr Chromium 51.996	25 Mn Manganese 54.9380	26 Fe Iron 55.847	27 Co Cobalt 58.9332	28 Ni Nickel 58.71	29 Cu Copper 63.546	30 Zn Zinc 65.37	31 Ga Gallium 69.72	32 Ge Germanium 72.59	33 As Arsenic 74.9216	34 Se Selenium 78.96	35 Br Bromine 79.904	36 Kr Krypton 83.80
37 Rb Rubidium 85.47	38 Sr Strontium 87.62	39 Y Yttrium 88.905	40 Zr Zirconium 91.22	41 Nb Niobium 92.906	42 Mo Molybdenum 95.94	43 Tc Technetium (99)	44 Ru Ruthenium 101.07	45 Rh Rhodium 102.905	46 Pd Palladium 106.4	47 Ag Silver 107.868	48 Cd Cadmium 112.40	49 In Indium 114.82	50 Sn Tin 118.69	51 Sb Antimony 121.75	52 Te Tellurium 127.60	53 I Iodine 126.9044	54 Xe Xenon 131.30
55 Cs Cesium 132.905	56 Ba Barium 137.34	57 La Lanthanum 138.91	72 Hf Hafnium 179.49	73 Ta Tantalum 180.948	74 W Tungsten 183.85	75 Re Rhenium 186.2	76 Os Osmium 190.2	77 Ir Iridium 192.2	78 Pt Platinum 195.09	79 Au Gold 196.967	80 Hg Mercury 200.59	81 Tl Thallium 204.37	82 Pb Lead 207.19	83 Bi Bismuth 208.980	84 Po Polonium (210)	85 At Astatine (210)	86 Rn Radon (222)
87 Fr Francium (223)	88 Ra Radium (226)	89 Ac Actinium (227)	104 Rf Rutherfordium (261)	105 Db Dubnium (262)	106 Sg Seaborgium (266)	107 Bh Bohrium (264)	108 Hs Hassium (269)	109 Mt Meitnerium (268)	110 Ds Darmstadtium (269)	111 Rg Roentgenium (272)	112 Uub Ununbium (277)	113 Uut §	114 Uuq Ununquadium (285)	115 Uup §	116 Uuh Ununhexium (289)	117 Uus §	118 Uuo Ununoctium (293)

Lanthanide Series

58 Ce Cerium 140.12	59 Pr Praseodymium 140.907	60 Nd Neodymium 144.24	61 Pm Promethium (145)	62 Sm Samarium 150.35	63 Eu Europium 151.96	64 Gd Gadolinium 157.25	65 Tb Terbium 158.924	66 Dy Dysprosium 162.50	67 Ho Holmium 164.930	68 Er Erbium 167.26	69 Tm Thulium 168.934	70 Yb Ytterbium 173.04	71 Lu Lutetium 174.97

Actinide Series

90 Th Thorium 232.038	91 Pa Protactinium (231)	92 U Uranium 238.03	93 Np Neptunium (237)	94 Pu Plutonium (242)	95 Am Americium (243)	96 Cm Curium (247)	97 Bk Berkelium (247)	98 Cf Californium (251)	99 Es Einsteinium (254)	100 Fm Fermium (257)	101 Md Mendelevium (258)	102 No Nobelium (259)	103 Lr Lawrencium (260)

§ Note: Elements 113, 115, and 117 are not known at this time but are included in the table to show their expected positions.

© John Wiley & Sons, Inc.

Figure 4-1: The periodic table of the elements.

Starting on the far left of the periodic table is Group IA, also known as the *alkali metals.* These most metallic of the elements are very reactive, meaning they tend to combine with other elements. They're found naturally in a bonded state, never in a pure state. Reactivity increases as you move down the periodic table. Group IIA is called the *alkaline earth metals.* Just like the alkali metals, the alkaline earth metals are highly reactive.

The large central block of the periodic table is occupied by the *transition metals,* which are mostly listed as Group B elements. Transition metals have properties that vary from extremely metallic, at the left side, to far less metallic, on the right side. The rightmost boundary of the metals is shaped like a staircase, shown in bold in Figure 4-1.

✔ Elements to the left of the staircase are *metals* (except hydrogen in Group IA).

✔ Elements to the right of the staircase are *nonmetals.*

✔ Elements bordering the staircase (boron, silicon, germanium, arsenic, antimony, tellurium, polonium, and astatine) are called *metalloids* because they have properties between those of metals and nonmetals. Chemists debate the membership of certain elements (especially polonium and astatine) within the metalloids, but the list here reflects an inclusive view. You can find these elements in Groups IIIA, IVA, VA, VIA, and VIIA.

Metals tend to be solid and shiny, to conduct electricity and heat, to give up electrons, and to be *malleable* (easily shaped) and *ductile* (easily drawn out into wire). Nonmetals have properties opposite those of metals. The most extreme nonmetals are the noble gases, in Group VIIIA on the far right of the table. The noble gases are *inert,* or extremely unreactive. One column to the left, in Group VIIA, is another key family of nonmetals, the *halogens.* In nature, the reactive halogens tend to bond with metals to form salts, such as sodium chloride (NaCl).

Be aware that some periodic tables use a different system for labeling the groups, in which each column is simply numbered from 1 to 18, left to right. Figure 4-1 shows both systems.

Elements within a group tend to have similar chemical properties because they have similar arrangements of electrons at their outermost borders or *shells* (we describe electron shells in greater detail in the next section). As you move across a period, however, the properties of the elements tend to change gradually. By the time you reach the right edge of the periodic table (Group VIIIA), the elements' outermost shells are completely full. With full outer shells, the elements are stable and have no drive to react. The Group VIIIA elements are called *noble gases* because they seem to consider themselves above the other elements. They're very unreactive and rarely bond to form compounds.

Periods (rows) 6 and 7 have an added wrinkle. Elements with atomic numbers 58 through 71 and 90 through 103 are lifted out of the regular order and placed below the rest of the table. These two series are the *lanthanides* and the *actinides,* respectively. These elements are separated from the table for two main reasons:

✔ Doing so prevents the table from being inconveniently wide.

✔ The lanthanides all have pretty similar properties, as do the actinides.

EXAMPLE

Q. The elements within a group have widely varying numbers of protons, neutrons, and electrons. Why, then, do elements in a group tend to have similar chemical properties?

A. Chemical properties come mostly from the arrangement of electrons in the outermost shell of an atom. Although the elements at the top and bottom of a given group (like fluorine and astatine, for example) have very different numbers of protons, neutrons, and electrons, the arrangements of electrons in their outermost shells are very similar.

---

1. Are the following elements metals or nonmetals?

a. selenium

b. fluorine

c. strontium

d. chromium

e. bismuth

Solve It

2. Why are the noble gases referred to as "noble"?

Solve It

Predicting Properties from Periodic and Group Trends

The whole point of the periodic table, aside from providing interior decoration for chemistry classrooms, is to help predict and explain the properties of the elements. These properties change as a function of the numbers of protons and electrons in the element.

Increasing numbers of protons increase the positive charge of the nucleus, which contributes to *electron affinity,* the attraction an atom has for an added electron. Electron affinity increases as move you up and to the right on the periodic table. Within a period, the more protons an element has, the stronger its electron affinity tends to be. This trend isn't perfectly smooth because other, more subtle factors are at work, but it's a good general description of what happens as you move across a period. *Note:* Group VIIIA elements on the far right of the table are an important exception. These elements have full outer shells and have no use for an added electron.

Increasing the number of electrons changes the reactivity of the element in predictable ways, based on how those electrons fill successive energy levels. Electrons in the highest energy level occupy the outermost shell of the atom and are called *valence electrons.* Valence electrons, which are involved in bonding (see Chapter 5), determine whether an element is reactive or unreactive. Because chemistry is really about the making and breaking of bonds, valence electrons are the most important particles for chemistry.

Atoms are most stable when their valence shells are completely filled with electrons. Chemistry happens because atoms attempt to fill the valence shells. Alkali metals in Group IA of the periodic table are so reactive because they need only to give up one electron to have a completely filled valence shell. Halogens in Group VIIA are so reactive because they need only to acquire one electron to have a completely filled valence shell. The elements within a group tend to have the same number of valence electrons and for that reason tend to have similar chemical properties. Elements in Groups IA and IIA tend to react strongly with elements in Group VIIA. Group B elements tend to react less strongly.

Elements in the A groups have the same number of valence electrons as the Roman numeral of their group. For example, magnesium in Group IIA has two valence electrons, carbon in Group IVA has four valence electrons, oxygen in Group VIA has six valence electrons, and so on.

In addition to reactivity, another property that varies across the table is *atomic radius,* or the geometric size (not the mass) of the atoms. As you move down or to the right on the periodic table, elements have both more protons and more electrons. However, only as you move down the table do the added electrons occupy higher energy levels. Therefore, the following occurs:

✔ Atomic radius tends to decrease as you move to the right, because the increasing positive charge of the nuclei pulls the electrons of that energy level inward.

✔ As you move down the table, atomic radius tends to increase, because even though you're adding positively charged protons to the nucleus, you're now adding electrons to higher and higher energy levels, which correspond to larger and larger electron shells.

So you can make good estimates about the relative sizes of the radii of different elements by comparing their placement in the periodic table.

Q. Chromium has more electrons than scandium. Why, then, does scandium have a larger atomic radius?

A. Even though chromium (Cr) has more electrons than scandium (Sc), those extra electrons occupy the same energy level, because the two elements are in the same row on the periodic table. Furthermore, chromium has more protons than scandium (as you can tell by chromium's position to the right of scandium), creating a more positively charged nucleus that pulls the electrons of a given energy level inward, thereby decreasing the atomic radius.

3. Which has a larger atomic radius, silicon or barium?

Solve It

4. Which has a stronger electron affinity, silicon or barium?

Solve It

5. How many valence electrons do the following elements have?

　　a. I

　　b. O

　　c. Ca

　　d. H

　　e. Ge

Solve It

6. Why are valence electrons important?

Solve It

Seeking Stability with Valence Electrons by Forming Ions

Elements are so insistent about having filled valence shells that they'll gain or lose valence electrons to do so. Atoms that gain or lose electrons in this way are called *ions*. You can predict what kind of ion many elements will form simply by looking at their position on the periodic table. With the exception of Row 1 (hydrogen and helium), all main group elements are most stable (think "happiest") with a full shell of eight valence electrons, known as an *octet*. Atoms tend to take the shortest path to a complete octet, whether that means ditching a few electrons to achieve a full octet at a lower energy level or grabbing extra electrons to complete the octet at their current energy level. In general, metals on the left side of the periodic table (and in the middle) tend to lose electrons, and nonmetals on the right tend to gain electrons.

You can predict just how many electrons an atom will gain or lose to become an ion. Group IA (alkali metal) elements lose one electron. Group IIA (alkaline earth metal) elements lose two electrons. Things get unpredictable in the transition metal region but follow a pattern once more with the nonmetals. Group VIIA (halogen) elements gain one electron. Group VIA elements tend to gain two electrons, and group VA elements tend to gain three electrons. In short, elements tend to lose or gain as many electrons as necessary to have valence shells resembling the elements in group VIIIA, the noble gases.

When atoms become ions, they lose the one-to-one balance between their protons and electrons and therefore acquire an overall charge:

   ✔ **Cations:** Atoms that lose electrons (like metals) acquire positive charge, becoming *cations,* such as Na^+ or Mg^{2+}.

   ✔ **Anions:** Atoms that gain electrons (like nonmetals) acquire negative charge, becoming *anions,* such as Cl^- or O^{2-}.

The superscripted numbers and signs in the atoms' symbols indicate the ion's overall charge. Cations have superscripts with + signs, and anions have superscripts with – signs. When the element sodium, Na, loses an electron, it loses one negative charge and is left with one overall positive charge because it now has one more proton than electron. So Na becomes Na^+.

Q. Fluorine and sodium are only two atomic numbers apart on the periodic table. Why, then, does fluorine form an anion, F^-, whereas sodium forms a cation, Na^+?

A. Fluorine (F) is in Group VIIA, just one group to the left of the noble gases, and therefore needs to gain only one electron to complete a valence octet. Sodium (Na) lies just one group to the right of the noble gases, having wrapped around into Group IA of the next energy level. Therefore, sodium needs to lose only one electron to achieve a full valence octet.

---

7. How many electrons will be gained or lost by the following elements when forming an ion?

   a. lithium

   b. selenium

   c. boron

   d. oxygen

   e. chlorine

Solve It

8. What type of ion is nitrogen most likely to form?

Solve It

9. What type of ion is beryllium most likely to form?

Solve It

Putting Electrons in Their Places: Electron Configurations

An awful lot of detail goes into determining just how many electrons an atom has (see the previous sections). The next step is to figure out where those electrons live. Several different schemes exist for depicting all this important information, but the *electron configuration* is a type of shorthand that captures much of the pertinent information.

Each numbered period (row) of the periodic table corresponds to a different *principal energy level,* with higher numbers indicating higher energy. Within each energy level, electrons can occupy different sublevels. Each sublevel is made up of different types of *orbitals.* Different types of orbitals have slightly different energy. Each orbital can hold up to two electrons, but electrons won't double up within an orbital unless no other open orbitals exist at the same energy level. Electrons fill up orbitals from the lowest energies to the highest.

There are four types of orbitals: *s, p, d,* and *f:*

- ✔ **s:** Period 1 consists of a single 1*s* orbital. A single electron in this orbital corresponds to the electron configuration of hydrogen, written as $1s^1$. The superscript written after the symbol for the orbital indicates how many electrons occupy that orbital. Filling the orbital with two electrons, $1s^2$ corresponds to the electron configuration of helium. Each higher principal energy level contains its own *s* orbital (2*s*, 3*s*, and so on), and these orbitals are the first to fill within those levels.

- ✔ **p:** In addition to *s* orbitals, principal energy levels 2 and higher contain *p* orbitals. There are three *p* orbitals at each level, accommodating a maximum of six electrons. Because the three *p* orbitals (also known as p_x, p_y, and p_z) have equal energy, they're each filled with a single electron before any receives a second electron. The elements in Periods 2 and 3 on the periodic table contain only *s* and *p* orbitals. The *p* orbitals of each energy level are filled only after the *s* orbital is filled.

✔ **d:** Period 4 and higher on the periodic table include *d* orbitals, of which there are five at each principal energy level, accommodating a maximum of ten electrons. The *d* orbital electrons are a major feature of the transition metals.

✔ **f:** Period 5 and higher include *f* orbitals, numbering seven at each level, accommodating a maximum of 14 electrons. The *f* orbital electrons are a hallmark of the lanthanides and the actinides (see the earlier section "Organizing the Periodic Table into Periods and Groups" for more on these rows).

Trying to visualize how electrons fill orbitals can get very confusing, so Figure 4-2 is a periodic table with the different orbitals put in place of the elements. It will help. A lot. To use the diagram, start at the upper left and read from left to right. When you reach the end of a row, go to the beginning of the next row down. Keep going until you reach your element.

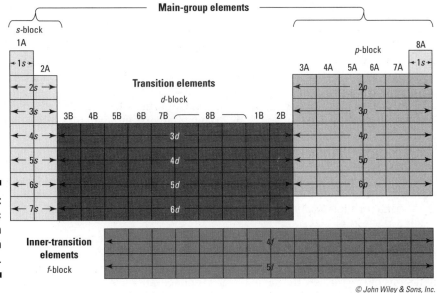

Figure 4-2: The periodic table with orbitals in place.

© John Wiley & Sons, Inc.

As you can see in Figure 4-2, after you get to Period 4, the exact order in which you fill the energy levels can get a bit confusing. To keep things straight, the Aufbau filling diagram in Figure 4-3 is useful. To use the diagram, start at the bottom and work your way up, from the lowest arrows to the highest. For example, always start by filling 1*s*, then fill 2*s*, then 2*p*, then 3*s*, then 4*s*, then 3*d*, and so on.

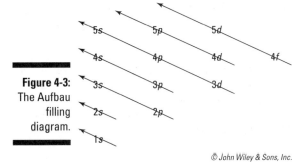

Figure 4-3: The Aufbau filling diagram.

© John Wiley & Sons, Inc.

Sadly, there are a few exceptions to the tidy picture presented by the Aufbau filling diagram. Copper, chromium, and palladium are notable examples (see Chapter 22 for details). Without going into teeth-grinding detail, these exceptional electron configurations arise from situations where electrons get transferred from their proper, Aufbau-filled orbitals to create half-filled or entirely filled sets of d orbitals; these half- and entirely filled states are more stable than the states produced by pure Aufbau-based filling.

To come up with a written electron configuration, you first determine how many electrons the atom in question actually has. Then you assign those electrons to orbitals, one electron at a time, from the lowest-energy orbitals to the highest. In a given type of orbital (like a $2p$ or $3d$ orbital, for example), you place only two electrons within the same orbital when there's no other choice. For example, suppose you want to find the electron configuration of carbon. Carbon has six electrons (the same as the number of its protons, as its atomic number indicates), and it's in Period 2 of the periodic table. First, the s orbital of level 1 is filled. Then, the s orbital of level 2 is filled. These orbitals each accept two electrons, leaving two more with which to fill the p orbitals of level 2. Each of the remaining electrons would occupy a separate p orbital. You wind up with $1s^2 2s^2 2p^2$. (Only at oxygen, $1s^2 2s^2 2p^4$, would electrons begin to double up in the $2p$ orbitals, indicated by the superscript 4 on the p.)

Electron configurations can get a bit long to write. For this reason, you may see them written in a condensed form, like $[Ne]3s^2 3p^3$. This condensed configuration is the one for phosphorus. The expanded electron configuration for phosphorus is $1s^2 2s^2 2p^6 3s^2 3p^3$. To abbreviate the configuration, simply go backward in atomic numbers until you hit the nearest noble gas (which, in the case of phosphorus, is neon). The symbol for that noble gas, placed within brackets, becomes the new starting point for the configuration. Next, simply include the standard configuration for those electrons beyond the noble gas (which, in the case of phosphorus, includes two electrons in $3s$ and three electrons in $3p$). The condensed form simply means that the atom's electron configuration is just like that of neon, with additional electrons filled into orbitals $3s$ and $3p$ as annotated.

Ions have different electron configurations than their parent atoms because ions are created by gaining or losing electrons. As you find out in the preceding section, atoms tend to gain or lose electrons so they can achieve full valence octets, like those of the noble gases. Guess what! The resulting ions have precisely the same electron configurations as those noble gases. For example, by forming the Br^- anion, bromine achieves the same electron configuration as the noble gas krypton.

Q. What is the electron configuration of titanium?

A. $1s^2 2s^2 2p^6 3s^2 3p^6 4s^2 3d^2$. Titanium (Ti) is atomic number 22 and therefore has 22 electrons to match its 22 protons. These electrons fill orbitals from lowest to highest energy in the order shown by the Aufbau filling diagram in Figure 4-3. Note that the $3d$ orbitals fill only after the $4s$ orbitals have filled, so titanium has two valence electrons.

10. What is the electron configuration of chlorine?

Solve It

11. What is the electron configuration of technetium?

Solve It

12. What is the condensed electron configuration of bromine?

Solve It

13. What are the electron configurations of a chlorine anion, an argon atom, and a potassium cation?

Solve It

Measuring the Amount of Energy (or Light) an Excited Electron Emits

Electrons can jump between energy levels by absorbing or releasing energy. When an electron absorbs an amount of energy exactly equivalent to the difference in energy between levels (a *quantum* of energy), the electron jumps to the higher energy level. This jump in energy creates an *excited state*. Excited states don't last forever; the lower-energy *ground state* is more stable. Excited electrons tend to release amounts of energy equivalent to the difference between energy levels and thereby return to ground state. The discrete particles of light that correspond to quanta of energy are called *photons*.

Light has properties of both a particle and a wave. A lovely result arises from light's properties: You can measure differences in electron energies simply by measuring the wavelengths of light emitted from excited atoms. In this way, you can identify different elements within a sample. The first basic relationship behind this technique is

$$\text{Speed of light } (c) = \text{Wavelength } (\lambda) \times \text{Frequency } (\nu)$$

where $c = 3.00 \times 10^8$ m/s. Frequency is often expressed in reciprocal seconds (1/s or s^{-1}), also known as *hertz* (1 Hz = 1 s^{-1}).

But what about energy? What's the relationship between frequency and energy? It must be ridiculously complicated, right? Wrong. The second basic relationship relating light to energy is

$$\text{Energy } (E) = \text{Planck's constant } (\hbar) \times \text{Frequency } (\nu)$$

where $\hbar = 6.626 \times 10^{-34}$ J·s. Here, J stands for *joules*, the SI unit of energy. Frequency is expressed in hertz (Hz), where 1 hertz is 1 inverse second (s^{-1}). So multiplying a frequency by Planck's constant, $\hbar$, yields *joules*, the units of energy.

Q. A hydrogen lamp emits blue light at a wavelength of 487 nanometers (nm). What is the frequency in hertz?

A. $6.16 \times 10^{14}\,s^{-1}$. Wavelength and the speed of light are known quantities here; you must solve for frequency, $\nu = c/\lambda$. Don't forget to convert nanometers to meters (1 nm = 10^{-9} m), because the speed of light is given in meters per second:

$$\nu = \frac{c}{\lambda} = \frac{(3.00 \times 10^8 \text{ m/s})}{(487 \times 10^{-9} \text{ m})} = 6.16 \times 10^{14} \text{ 1/s}$$

Q. What is the energy of emitted light with a frequency of 6.88×10^{14} Hz?

A. 4.56×10^{-19} J. You must recall here that Hz = s^{-1}.

$$E = \hbar\nu = (6.626 \times 10^{-34} \text{ J·s})(6.88 \times 10^{14} \text{ 1/s})$$
$$= 4.56 \times 10^{-19} \frac{\text{J·s}}{\text{s}} = 4.56 \times 10^{-19} \text{ J}$$

14. What is the frequency of a beam of light with wavelength 2.57×10^2 m?

Solve It

15. One useful, high-energy emission from excited electrons is the X-ray. What is the wavelength (in meters) of an X-ray with a frequency of $\nu = 8.72 \times 10^{17}$ Hz?

Solve It

16. What is the energy of a wave if it has a frequency of 3.91×10^8 Hz?

Solve It

Answers to Questions on the Periodic Table

Here are the answers to the practice problems in this chapter. Pat yourself on the back, whether in congratulation or consolation.

1 Metals are found to the left of the staircase, with the exception of hydrogen, and nonmetals are found to the right of the staircase. Several elements adjoining the staircase are classified as metalloids.

a. selenium (Se) = **nonmetal**

b. fluorine (F) = **nonmetal**

c. strontium (Sr) = **metal**

d. chromium (Cr) = **metal**

e. bismuth (Bi) = **metal**

2 The noble gases are described as "noble" because they seem to consider it beneath themselves to react with other elements. Because these elements have completely filled valence shells, they have no energetic reason to react. They're already as stable as they can be.

3 **Barium's atomic radius is larger.** Atomic size tends to increase from top to bottom and from right to left. Barium (Ba) is farther down and to the left on the periodic table than silicon (Si). Barium has more energy levels and is therefore larger.

4 **Silicon has stronger electron affinity.** Electron affinity tends to increase from left to right; silicon is to the right of barium on the period table.

5 When you're looking at elements in A groups, the number of valence electrons matches the element's group number.

a. Iodine (I) has **seven valence electrons** because it's in Group VIIA.

b. Oxygen (O) has **six valence electrons** because it's in Group VIA.

c. Calcium (Ca) has **two valence electrons** because it's in Group IIA.

d. Hydrogen (H) has **one valence electron** because it's in Group IA.

e. Germanium (Ge) has **four valence electrons** because it's in Group IVA.

6 Valence electrons are important because they occupy the highest energy level in the outermost shell of the atom. As a result, valence electrons are the electrons that see the most action, as in forming bonds or being gained or lost to form ions. The number of valence electrons in an atom largely determines the chemical reactivity of that atom and many other properties.

7 Elements seek stable valence shells by gaining or losing electrons. Whether an element gains or loses electrons has to do with where that element sits in the periodic table. For elements on the left side, losing just a few electrons is easier than gaining many more; for these elements, the Roman numeral of the A group to which they belong tells you how many electrons they lose — and therefore tells you their positive charge. For elements on the right side of the table, gaining a few electrons is easier than losing many more; these elements gain the number of electrons necessary to create valence shells like those in Group VIIIA (the noble gases). An element in Group VIA therefore gains two electrons.

a. Lithium (Li) loses one electron, forming the lithium cation Li$^+$.

b. Selenium (Se) gains two electrons to form the selenide anion Se^{2-}.

c. Boron (B) loses three electrons to form the boron cation B^{3+}.

d. Oxygen (O) gains two electrons to form the oxide anion O^{2-}.

e. Chlorine (Cl) gains one electron to form the chloride anion Cl$^-$.

8 **N^{3-},** also known as *nitride* or the *nitrogen anion*. By gaining three electrons, nitrogen, which is in Group VA, assumes a full octet, like neon.

9 **Be^{2+},** also known as the *beryllium cation*. By losing two electrons, beryllium, which is in Group IIA, assumes a full octet valence shell, with two electrons like helium.

10 **$1s^2 2s^2 2p^6 3s^2 3p^5$.** Chlorine (Cl) has 17 electrons.

11 **$1s^2 2s^2 2p^6 3s^2 3p^6 4s^2 3d^{10} 4p^6 5s^2 4d^5$.** Technetium (Tc) has 43 electrons, which fill according to the Aufbau diagram.

12 **[Ar]$4s^2 3d^{10} 4p^5$.** Condense the configuration for bromine (Br) by summarizing all the previous entirely filled rows with a single noble gas in brackets. The expanded electron configuration of bromine is $1s^2 2s^2 2p^6 3s^2 3p^6 4s^2 3d^{10} 4p^5$. Bromine is in Row 4, so consolidate the electron configurations of Periods 1 through 3 as [Ar], because argon is the last element in Row 3. Then add to [Ar] the configuration of the remaining electrons, which fill $4s^2 3d^{10} 4p^5$.

13 **All three have the configuration $1s^2 2s^2 2p^6 3s^2 3p^6$,** the configuration of the noble gas argon (Ar). Both chlorine (Cl) and potassium (K) seek stable, full octet configurations by forming ions and achieve the configuration of argon by gaining or losing valence electrons.

14 **$1.17 \times 10^6 \, s^{-1}$.**

$$v = \frac{c}{\lambda} = \frac{(3.0 \times 10^8 \text{ m/s})}{(2.57 \times 10^2 \text{ m})} = 1.17 \times 10^6 \, 1/s$$

15 **3.44×10^{-10} m.**

$$\lambda = \frac{c}{v} = \frac{(3.0 \times 10^8 \text{ m/s})}{(8.72 \times 10^{17} \, 1/s)} = 3.44 \times 10^{-10} \text{m}$$

16 **2.59×10^{-25} J.**

$$E = \hbar v = (6.626 \times 10^{-34} \text{ J} \cdot \text{s})(3.91 \times 10^8 \, 1/s) = 2.59 \times 10^{-25} \text{ J}$$

Part II
Making and Remaking Compounds

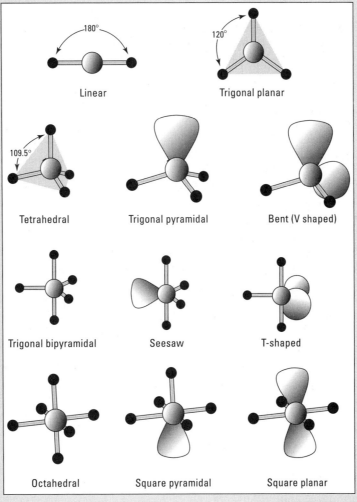

Linear

Trigonal planar

Tetrahedral

Trigonal pyramidal

Bent (V shaped)

Trigonal bipyramidal

Seesaw

T-shaped

Octahedral

Square pyramidal

Square planar

© John Wiley & Sons, Inc.

web
extras

Get details on naming compounds, converting to and from moles, and more on the Cheat Sheet at
www.dummies.com/extras/chemistrywb.

In this part . . .

✔ Atoms form bonds with each other based on a variety of factors determined by their need for electrons. You can depict these structures by drawing Lewis structures to determine polarity and geometry.

✔ Every syllable in the name of a chemical compound conveys something about that compound. From ionic and molecular (covalent) compounds to organic hydrocarbons and acids, the names matter, and you find out why in this part.

✔ The mole is a unit that helps chemists relate the incredibly small size of an atom to the incredibly large number of atoms that chemists must deal with on a daily basis. Meet the mighty mole in this part.

✔ Chemical reactions act as a recipe for a chemist to determine what will actually happen when two or more compounds are mixed together. Reactions require balancing and sometimes won't even take place, depending on their net ionic result.

✔ Beyond chemical reactions lies the strange and wonderful concept of stoichiometry. Stoichiometry allows scientists to calculate quantitative (numerical) amounts of reactants and products based on balanced chemical reactions. In short (and in plain English), this part tells you how much of something a reaction will make.

Chapter 5

Building Bonds

In This Chapter

▶ Giving and receiving electrons in ionic bonding

▶ Sharing electrons in covalent bonding

▶ Understanding molecular orbitals

▶ Tugging at the idea of polarity

▶ Shaping up molecules with VSEPR theory and hybridization

Many atoms are prone to public displays of affection, pressing themselves against other atoms in an intimate electronic embrace called *bonding.* Atoms bond with one another by playing various games with their valence electrons. In this chapter, we describe the basic rules of those games.

Because valence electrons are so important to bonding, chemistry problems involving bonding sometimes make use of *electron dot structures,* symbols that represent valence electrons as dots surrounding an atom's chemical symbol. You should be able to draw and interpret electron dot structures for atoms, as in Figure 5-1. This figure shows the electron dot structures for elements in the periodic table's first two rows. Notice that the valence shells progressively fill moving from left to right.

 To determine the electron dot structure of any element, count the number of electrons in that element's valence shell. Then draw that number of dots around the chemical symbol for the element. To do so, imagine the chemical symbol as a square. Start from the top of the symbol and, going clockwise, put one dot on each side until you run out of valence electrons. Don't double up on any side until you've gone around the square once.

Chapter 4 describes some of the factors that determine whether atoms gain or lose electrons to form ions. Make sure you understand those patterns before attacking this chapter.

Figure 5-1:
Electron dot structures for elements in the first two rows of the periodic table.

IA	IIA	IIIA	IVA	VA	VIA	VIIA	VIIIA
H ·							He :
Li ·	·Be·	·Ḃ·	·Ċ·	:N̈:	:Ö:	:F̈:	:N̈e:

© John Wiley & Sons, Inc.

Pairing Charges with Ionic Bonds

Atoms of some elements, like metals, can easily lose valence electrons to form *cations* (atoms with positive charge) that have stable electron configurations. Atoms of other elements, like the halogens, can easily gain valence electrons to form *anions* (atoms with negative charge) with stable electron configurations. Cations and anions experience *electrostatic attraction* to one another due to the opposite charges present. So a cation will snuggle up to an anion, given the chance. The attraction between cations and anions is called *ionic bonding,* and it happens because the energy of the ionically bonded ions is lower than the energy of the ions when they're separated.

You can think of an ionic bond as resulting from the transfer of an electron from one atom to another, as Figure 5-2 shows for sodium and chlorine. Metals (like sodium) tend to give up their electrons to nonmetals (like chlorine) because nonmetals are much more *electronegative* (they more strongly attract electrons within a bond to themselves). The greater the difference in electronegativity between the two ions, the more ionic the bond that forms between them (where *ionic* means completely uneven in the sharing of electrons). So in short, ionic bonding is generally a transfer of electrons from a metal to a nonmetal.

Figure 5-2:
The transfer
of an
electron
from sodium
to chlorine
to form an
ionic bond
between the
Na$^+$ cation
and the
Cl$^-$ anion.

$$Na \bullet \overset{\frown}{\longrightarrow} \bullet \ddot{\underset{\cdot\cdot}{Cl}} \colon \longrightarrow Na^+ \colon \ddot{\underset{\cdot\cdot}{Cl}} \colon^-$$

© John Wiley & Sons, Inc.

Although ions are often individual charged atoms, there are many examples of *polyatomic ions,* which are charged particles made up of more than one atom. Common polyatomic ions include ammonium, NH_4^+, and sulfate, SO_4^{2-}.

When cations and anions associate in ionic bonds, they form *ionic compounds.* At room temperature, most ionic compounds are solid because of the strong electrostatic forces that hold together the ions within them. The ions in ionic solids pack together in a *lattice,* a highly organized, regular arrangement that allows for the maximum electrostatic interaction while reducing the repulsive forces between anions and between cations. The geometric details of the packing can differ among different ionic compounds, but you can see a simple lattice structure in Figure 5-3. Flip to Chapter 6 for full details on polyatomic ions and ionic compounds.

Figure 5-3:
The lattice structure of an ionic solid, sodium chloride.

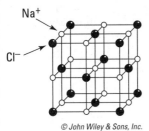

Na$^+$

Cl$^-$

© John Wiley & Sons, Inc.

The strong electrostatic forces that hold together ionic lattices result in the high melting and boiling points that are common among ionic compounds (see Chapter 10 for general information on melting and boiling points). Although disrupting ionic bonds may take a great deal of thermal energy, many ionic compounds are easily dissolved in water or in other *polar solvents* (fluids made up of molecules that have unevenly distributed charge).

When the solvent molecules are polar, they can engage in favorable interactions with the ions, helping to compensate for disrupting the ionic bonds. For example, polar water molecules can interact well with both sodium cations (Na$^+$) and chlorine anions (Cl$^-$). Water molecules are polar because they have distinct and separate bits of positive and negative charge. Water molecules can orient their positive bits toward Cl$^-$ and their negative bits toward Na$^+$. Positive charges attract negative charges and vice versa, so these kinds of interactions are favorable; that is, they require less energy. So water dissolves solid NaCl quite well because the water-ion interactions can compete with the Na$^+$–Cl$^-$ interactions.

When ionic compounds are melted or dissolved so that the individual ions can move about, the resulting liquid is a very good conductor of electricity. Ionic solids, however, are often poor conductors of electricity.

Salts are a common variety of ionic compound. A salt can form from the reaction between a base and an acid (both of which we cover in Chapter 16). For example, hydrochloric acid reacts with sodium hydroxide to form sodium chloride (a salt) and water:

$$HCl(aq) + NaOH(aq) \rightarrow NaCl(aq) + H_2O(l)$$

Note that *aq* indicates that the substance is dissolved in water, in an *aqueous* solution.

EXAMPLE

Q. Why do metals tend to form ionic compounds with nonmetals?

A. Metals are much less electronegative than nonmetals, meaning that they give up valence electrons much more easily. Nonmetals (especially Group VIIA and VIA nonmetals) very easily gain new valence electrons. So metals and nonmetals tend to form bonds in which the metal atoms entirely surrender valence electrons to the nonmetals. Bonds with extremely unequal electron-sharing are called *ionic bonds*.

1. Draw the electron dot structure of potassium fluoride and show how the electron is transferred between the elements potassium and fluorine to create the ionic compound.

Solve It

2. The ionic compound lithium sulfide forms between the elements lithium and sulfur. In which direction are electrons transferred to form ionic bonds, and how many electrons are transferred?

Solve It

3. Magnesium chloride is dissolved in a beaker of water and in a beaker of rubbing alcohol until no more compound will dissolve. An electrical circuit is set up for each beaker, with wires leading from a battery into the solution and a separate set of wires leading from the solution to a light bulb. The bulb connected to the aqueous solution circuit glows more brightly than the bulb connected to the alcohol solution circuit. Why?

Solve It

Sharing Electrons with Covalent Bonds

REMEMBER

Sometimes the way for atoms to reach their most stable, lowest-energy states is to share valence electrons. When atoms share valence electrons, chemists say that they're engaged in *covalent bonding.* The very word *covalent* means "together in valence." Compared to ionic bonding, covalent bonding tends to occur between atoms of similar electronegativity, especially between nonmetals.

Just as ionic bonds tend to form in such a way that both atoms end up with completely filled valence shells, the atoms involved in covalent bonds tend to share electrons in such a way that each ends up with a completely filled valence shell. The shared electrons are attracted to the nuclei of both atoms, forming the bond.

The simplest and best-studied covalent bond is the one that forms between two hydrogen atoms (see Figure 5-4). Separately, each atom has only one electron with which to fill its $1s$ orbital. By forming a covalent bond, each atom lays claim to two electrons within the molecule of dihydrogen (H_2). The figure shows various ways in which a covalent bond can be represented: by explicitly depicting the valence shells (a), by using electron dot structures (b), and by signifying a shared pair of electrons with a single line (c). The latter two ways to show bonding are referred to as *Lewis structures.*

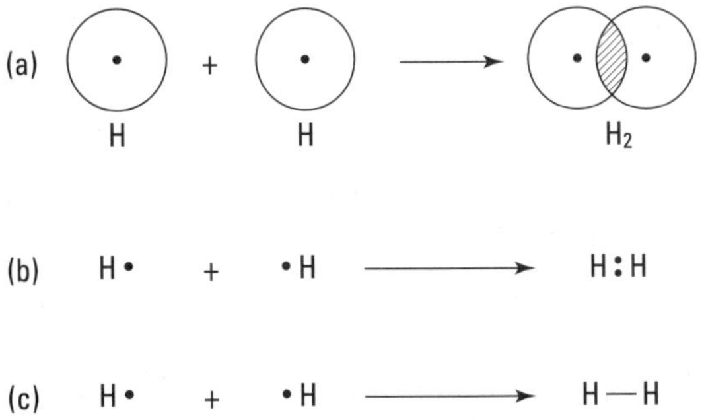

Figure 5-4:
Three representations of the formation of a covalent bond in a dihydrogen molecule.

(a)   H   +   H $\longrightarrow$ H_2

(b)   H•   +   •H $\longrightarrow$ H:H

(c)   H•   +   •H $\longrightarrow$ H—H

© John Wiley & Sons, Inc.

Atoms can share more than a single pair of electrons. When atoms share two pairs of electrons, they're said to form a *double bond,* and when they share three pairs of electrons, they're said to form a *triple bond.* Figure 5-5 shows examples of double and triple bonds using electron dot structures and line structures.

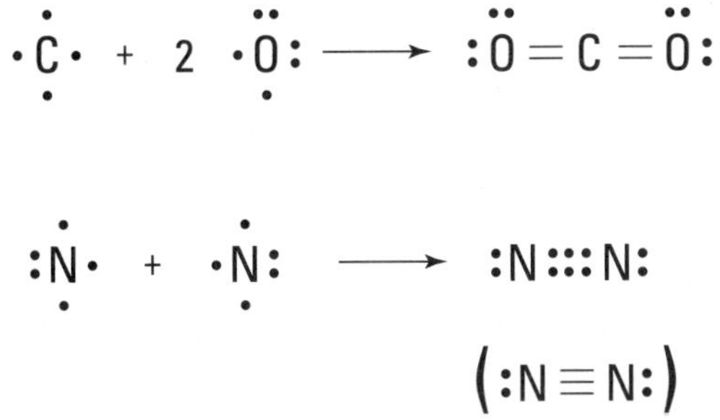

Figure 5-5:
The formation of double bonds in carbon dioxide and triple bonds in dinitrogen.

•C• + 2 •Ö: $\longrightarrow$:Ö=C=Ö:

:N• + •N: $\longrightarrow$:N:::N:

$\left(:N\equiv N:\right)$

© John Wiley & Sons, Inc.

If you know a molecule's formula, a few guidelines can help you figure out a correct Lewis structure for the molecule. In this example, we work out the Lewis structure of formaldehyde, CH_2O (Figure 5-6 can help you follow along):

1. **Add up all the valence electrons for all the atoms in the molecule.**

   The valence electrons are the ones you can use to build the structure. Account for any extra or missing electrons in the case of ions. For example, if you know your molecule has a +2 charge, remember to subtract 2 from the total number of valence electrons. If your molecule has a –2 charge, remember to add 2 to the total valence number.

   In the case of formaldehyde, C has four valence electrons, each H has one valence electron, and O has six valence electrons. The total number of valence electrons is 12.

2. **Pick a "central" atom to serve as the anchor of your Lewis structure.**

   The central atom is usually the one that can form the most bonds, which is often the atom with the most empty valence orbital slots to fill. For larger molecules, this step may involve some trial-and-error, but for smaller molecules, some choices are obviously better than others. For example:

   • Carbon is usually a very good central atom because it can form four bonds.

   • Hydrogen will never be the central atom for any compound because it can form only one bond.

   In the case of formaldehyde, carbon is the obvious first choice because of the four bonds it can form. Oxygen can form only two bonds, and hydrogen can form only one bond, so neither is the central atom in this molecule.

3. **Connect the other, "outer" atoms to your central atom using single bonds only.**

   Each single bond counts for two electrons. In the case of formaldehyde, attach the single oxygen and each of the two hydrogen atoms to the central carbon atom.

4. **Fill the valence shells of your outer atoms. Then put any remaining electrons on the central atom.**

   In our example, carbon and oxygen should each have eight electrons in their valence shells; each hydrogen atom should have two. However, by the time you fill the valence shells of the outer atoms (oxygen and the two hydrogens), you've used up your allotment of 12 electrons.

5. **Check whether the central atom now has a full valence shell.**

   If the central atom has a full valence shell, then your Lewis structure is drawn properly — it's formally correct even though it may not correspond to a real structure. If the central atom still has an incompletely filled valence shell, then take electron dots (nonbonding electrons) from outer atoms and use them to create double and/or triple bonds to the central atom until the central atom's valence shell is filled.

   Remember, each added bond requires two more electrons. A single bond involves two shared electrons, a double bond involves four shared electrons, and a triple bond involves six shared electrons.

In the case of the formaldehyde molecule, you have to create a double bond between carbon and one of the outer atoms. Oxygen is the only choice for a double-bond partner, because each hydrogen can accommodate only two electrons in its shell. So use two of the electrons assigned to oxygen to create a second bond with carbon.

1. $C(4\ e^-) + H(1\ e^-) + H(1\ e^-) + O(6\ e^-) = 12\ e^-$

2. Carbon is central atom; it can form more bonds (4) than O, H.

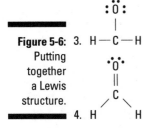

Figure 5-6: Putting together a Lewis structure.

© John Wiley & Sons, Inc.

Sometimes a covalent bond forms when one atom donates both electrons to the bond, with the other atom contributing no electrons. This kind of bond is called a *coordinate covalent bond*. Atoms with lone pairs are capable of donating both electrons to a coordinate covalent bond. A *lone pair* consists of two electrons paired within the same orbital that aren't used in bonding. Even though covalent bonding usually occurs between nonmetals, metals can engage in coordinate covalent bonding. Usually, the metal receives electrons from an electron donor called a *ligand*.

Sometimes a given set of atoms can covalently bond with each other in multiple ways to form a compound. This situation leads to something called *resonance*. Each of the possible bonded structures is called a *resonance structure*. The actual structure of the compound is a *resonance hybrid*, a sort of weighted average of all the resonance structures. For example, if two atoms are connected by a single bond in one resonance structure and the same two atoms are connected by a double bond in a second resonance structure, then in the resonance hybrid, those atoms are connected by a bond that is worth 1.5 bonds. A common example of resonance is found in ozone, O_3, shown in Figure 5-7.

Figure 5-7: Two representations of resonance structures of ozone.

© John Wiley & Sons, Inc.

Q. Draw a Lewis structure for propene, C_3H_6.

A. First, add up the total valence electrons. Each carbon atom contributes 4 electrons, and each hydrogen contributes 1, for a total of 18 valence electrons. Next, pick a central atom. The only choice available is carbon, because hydrogen can only have one bond and thus can never be a central atom in a Lewis structure. You have three carbons, so just connect the three into a carbon chain. In formulas containing just carbon and hydrogen, you'll always find the carbons bonded together in a chain with the hydrogens bonded around the outside.

With the three carbons connected together, you've used up four of the total valence electrons, because each bond counts as two electrons. You then can place the hydrogens around the carbons as you see fit, as long as you remember that each carbon can only have four total bonds or lone electron pairs. After placing the hydrogens, you end up with eight total bonds in the structure for a total of 16 valence electrons. This leaves two electrons that you can place on a carbon atom that does not have four total bonds or electron pairs present.

You're now finished with the simple part. One carbon atom in the structure still requires two additional electrons to fill its valence shell. The only way to fill this shell is to take the lone pair of electrons you added to one of the carbons and instead use it to create a double bond between two of the carbons. You then need to move the hydrogens around to ensure that each carbon has a total of four bonds. Only one arrangement of hydrogen atoms to the three carbons allows you to fill all the carbon valence shells, as you can see in the following figure:

$$H-\overset{\overset{\displaystyle H}{|}}{\underset{\underset{\displaystyle H}{|}}{C}}-\overset{\overset{\displaystyle H}{|}}{C}=C\overset{\diagup H}{\diagdown H}$$

© John Wiley & Sons, Inc.

4. Bertholite is the common name for dichlorine (Cl_2), a toxic gas that has been used as a chemical weapon. Why is bertholite most certainly a covalently bonded compound? What is the most likely electron dot structure of this compound?

Solve It

5. When aluminum chloride salt is dissolved in water, aluminum (III) cations become surrounded by clusters of six water molecules to form a "hexahydrated" aluminum cation, $Al(H_2O)_6^{3+}$. Being a Group IIIA metal, aluminum easily gives up its valence electrons. The oxygen atom in water possesses two lone pairs. What kind of bonding most likely occurs between the aluminum and the hydrating water molecules?

Solve It

6. Benzene, C_6H_6, is a common industrial solvent. The benzene molecule is based on a ring of covalently bonded carbon atoms. Draw two acceptable Lewis structures for benzene. Based on the structures, describe a likely resonance hybrid structure for benzene.

Solve It

Occupying and Overlapping Molecular Orbitals

Electrons occupy distinct orbitals within atoms (see Chapter 4 for details). When atoms covalently bond to form molecules, the shared electrons are no longer constrained to those atomic orbitals; instead, they occupy *molecular orbitals,* larger regions that form from the overlap of atomic orbitals. Just as different atomic orbitals are associated with different levels of energy, so are molecular orbitals. A stable covalent bond forms between two atoms because the energy of the molecular orbital associated with the bond is lower than the combined energies associated with the atomic orbitals of the separated atoms.

Because electrons have wave-like properties, atomic orbitals can be described with wave functions and can overlap in different ways:

- ✔ **Low-energy bonding orbital:** In one mode, the orbitals interact *favorably* (with low energy) and together occupy a *bonding orbital.*

- ✔ **Higher-energy antibonding orbital:** In another mode, the orbitals interact unfavorably within a higher-energy *antibonding orbital.*

The energy relationships between unbound atoms and different types of molecular orbitals are summarized in *molecular orbital diagrams,* such as the one for hydrogen in Figure 5-8. In this figure, two hydrogen atoms each contribute a single electron from a $1s$ orbital to a sigma (σ) bonding orbital. The low-energy bonding orbital is favored over the higher-energy sigma antibonding (σ^*) orbital. This illustrates a general principle: Given a choice between high- and low-energy states, molecules prefer the low-energy states. This preference for lower energy is what is meant by *favorable* (low energy) versus *unfavorable* (high energy).

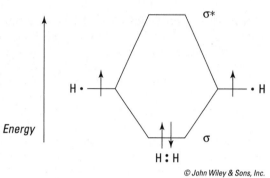

Figure 5-8:
A molecular diagram for the formation of a hydrogen molecule.

Energy

© John Wiley & Sons, Inc.

REMEMBER

In addition to differences in the interaction of orbitals, covalent bonds can differ based on the shape of the molecular orbitals:

✓ **Sigma bond:** When atomic orbitals overlap head to head in such a way that the resulting molecular orbital is symmetric with respect to the *bond axis* (the line connecting the two bonded atoms), chemists say that a σ bond *(sigma bond)* is formed.

✓ **Pi bond:** When atomic orbitals overlap side to side in such a way that the resulting molecular orbital is symmetric with the bond axis in only one plane, chemists say that a π bond *(pi bond)* is formed.

Sigma bonds are stronger than pi bonds because the electrons within sigma bonds lie directly between the two atomic nuclei. The negatively charged electrons in sigma bonds therefore experience favorable (as in low-energy) attraction to the positively charged nuclei. Electrons in pi bonds are farther away from the nuclei, so they experience weaker attraction.

REMEMBER

Sigma bonds form when *s* or *p* orbitals overlap in a head-on manner. Single bonds are usually sigma bonds. Pi bonds are usually double or triple bonds. Figure 5-9 depicts these situations.

Figure 5-9:
Formation of a sigma bond (σ) from two *s* orbitals and formation of a pi bond (π) from two adjacent *p* orbitals.

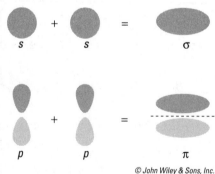

© John Wiley & Sons, Inc.

Q. Both sigma bonds and pi bonds have a kind of symmetry with respect to the two atoms in the bond. What is the difference in a sigma bond's symmetry versus a pi bond's symmetry?

A. In any bond between two atoms, you can imagine an imaginary line (the bond axis) that connects the center of one atom to the center of the other atom. Sigma bonds are perfectly symmetric about this line; you can imagine the sigma bond as a kind of tube that wraps around the bond axis. If you were to rotate the bonded atoms around the imaginary line, the bond would look the same all the way around.

Pi bonds, on the other hand, are symmetric to the bond axis *in only one plane.* You can imagine the two atoms pressed onto a flat surface; the bond axis is the imaginary line on this surface. The pi bonds connect the two atoms above the line and below the line. If you were to rotate the bonded atoms around the imaginary line, the pi bonds would rise up off the surface and sink below the surface as you rotated them, like the planks on a paddlewheel rise above the surface of the water and then sink below the surface of the water.

7. Draw a molecular orbital diagram for the hypothetical molecule dihelium (He_2).

Solve It

8. Based on the molecular orbital diagram of dihelium (He_2), explain why dihelium is far less likely to exist than hydrogen (H_2).

Solve It

9. Double bonds involve one sigma bond and one pi bond. A simple molecule that contains a double bond is ethene, $H_2C=CH_2$. Ethene reacts with water to form ethanol:

$$H_2C=CH_2+H_2O \rightarrow H_3C-COH$$

This reaction is favorable, meaning that it progresses on its own, without any input of energy. Why might this be the case?

Solve It

Polarity: Sharing Electrons Unevenly

The earlier sections on ionic bonding and covalent bonding refer to the concept of *electronegativity,* or the tendency of an atom to draw electrons toward itself. Ionic bonds form between atoms with large differences in electronegativity, whereas covalent bonds form between atoms with smaller differences in electronegativity. In truth, there's no natural distinction between the two types of bonds (ionic and covalent); they lie on opposite sides of a spectrum of *polarity,* or unevenness in the distribution of electrons within a bond.

The greater the difference in electronegativity between two atoms, the more polar the bond is that forms between them. Imagine the electrons in the bond as being spread out into a cloud within the molecular orbital. In polar bonds, the cloud is denser in the vicinity of the more electronegative atom. In nonpolar bonds, like those formed between atoms of the same element, the cloud is evenly distributed between both atoms. Polar bonds have more ionic character, whereas nonpolar bonds have more covalent character. Here's how to distinguish the character of a bond:

- ✔ Usually, a difference in electronegativity less than about 0.4 means that the corresponding covalent bond is considered nonpolar. This indicates that both atoms have nearly an equal pull for the electrons shared between them, so neither atom ends up with the electrons more than the other. Nonpolar bonds are most commonly found in diatomic molecules.

- ✔ Differences in electronegativity ranging from 0.4 to about 1.9 correspond to increasingly polar covalent bonds.

- ✔ Above about 1.9, the bond is increasingly considered ionic.

Figure 5-10 displays the electronegativities of the elements to help clarify why atoms that lie farther from each other horizontally on the periodic table tend to form more polar bonds.

Decreasing ↓

Increasing →

1 H 2.1																
3 Li 1.0	4 Be 1.5											5 B 2.0	6 C 2.5	7 N 3.0	8 O 3.5	9 F 4.0
11 Na 0.9	12 Mg 1.2											13 Al 1.5	14 Si 1.8	15 P 2.1	16 S 2.5	17 Cl 3.0
19 K 0.8	20 Ca 1.0	21 Sc 1.3	22 Ti 1.5	23 V 1.6	24 Cr 1.6	25 Mn 1.5	26 Fe 1.8	27 Co 1.9	28 Ni 1.9	29 Cu 1.9	30 Zn 1.6	31 Ga 1.6	32 Ge 1.8	33 As 2.0	34 Se 2.4	35 Br 2.8
37 Rb 0.8	38 Sr 1.0	39 Y 1.2	40 Zr 1.4	41 Nb 1.6	42 Mo 1.8	43 Tc 1.9	44 Ru 2.2	45 Rh 2.2	46 Pd 2.2	47 Ag 1.9	48 Cd 1.7	49 In 1.7	50 Sn 1.8	51 Sb 1.9	52 Te 2.1	53 I 2.5
55 Cs 0.7	56 Ba 0.9	57 La 1.1	72 Hf 1.3	73 Ta 1.5	74 W 1.7	75 Re 1.9	76 Os 2.2	77 Ir 2.2	78 Pt 2.2	79 Au 2.4	80 Hg 1.9	81 Tl 1.8	82 Pb 1.9	83 Bi 1.9	84 Po 2.0	85 At 2.2
87 Fr 0.7	88 Ra 0.9	89 Ac 1.1														

Electronegativities of the Elements

© John Wiley & Sons, Inc.

Figure 5-10: Electronegativities of the elements.

Differences in the electronegativity of atoms create polarity in bonds between those atoms. Because electrons are distributed unevenly within a polar covalent bond, the more electronegative atom takes on a partial negative charge, signified by the symbol δ^-. The less electronegative atom takes on a partial positive charge, δ^+. This difference in charge along the axis of the bond is called a *dipole*. The polar bonds within a molecule add to create polarity in the molecule as a whole. The precise way in which the individual bonds contribute to the overall polarity of the molecule depends on the shape of the molecule:

- ✔ **Two very polar bonds point in opposite directions:** Their polarities cancel out.

- ✔ **Two polar bonds point in the same direction:** Their polarities add.

- ✔ **Two polar bonds point at each other so they're diagonal to one another:** Their polarities cancel in one direction but add in a perpendicular direction.

Individual bonds have dipoles, which sum over all the bonds of a molecule (taking geometry into account) to create a *molecular dipole*. In addition to the permanent dipoles created by polar bonds, instantaneous dipoles can be temporarily created within nonpolar bonds and molecules. Both kinds of dipoles play important roles in the ways molecules interact:

- ✔ Permanent dipoles lead to dipole-dipole interactions and to hydrogen bonds.

- ✔ Instantaneous dipoles lead to attractive London forces, which affect molecules in solutions.

Q. Why doesn't it make sense to ask whether an element (like hydrogen or fluorine) engages in polar bonds versus nonpolar bonds?

A. You know whether an element engages in polar or nonpolar bonds only with respect to specific bonds with other elements. For example, hydrogen engages in a perfectly nonpolar covalent bond with another hydrogen atom in the molecule H_2. Likewise, fluorine engages in a nonpolar bond with another fluorine atom in F_2. On the other hand, the bond between hydrogen and fluorine in the compound HF is very polar because of the large electronegativity difference the two atoms.

10. Predict whether bonds between the following pairs of atoms are nonpolar covalent, polar covalent, or ionic:

a. H and Cl

b. Ga and Ge

c. O and O

d. Na and Cl

e. C and O

Solve It

11. Tetrafluoromethane (CF_4) contains four covalent bonds. Water (H_2O) contains two covalent bonds. Which molecule has bonds with more polar character?

Solve It

Shaping Molecules: VSEPR Theory and Hybridization

We'll start with the hard part: VSEPR stands for *valence shell electron pair repulsion*. Okay, now it gets easier. VSEPR is simply a model that helps predict and explain why molecules have the shapes they do. Molecular shapes help determine how molecules interact with each other. For example, molecules that stack nicely on one another are more likely to form solids. And two molecules that can fit together so their reactive bits lie closer together in space are more likely to react with one another.

The basic principle underlying VSEPR theory is that valence electron pairs, whether they're lone pairs or they occur within bonds, prefer to be as far from one another as possible. There's no sense in crowding negative charges any more than necessary.

Of course, when multiple pairs of electrons participate in double or triple covalent bonds, those electrons stay within the same bonding axis. Lone pairs repel other lone pairs more strongly than they repel bonding pairs, and the weakest repulsion is between two pairs of bonding electrons. Two lone pairs separate themselves as far apart as they can go, on exact opposite sides of an atom if possible. Electrons involved in bonds also separate themselves as far apart as they can go but with less force than two lone pairs. In general, all electron pairs try to maintain the maximum mutual separation. But when an atom is bonded to many other atoms, the "ideal" of maximum separation isn't always possible because bulky groups bump into one another. So the final shape of a molecule emerges from a kind of negotiation between competing interests.

VSEPR theory predicts several shapes that appear over and over in real-life molecules. You can see them in Figure 5-11.

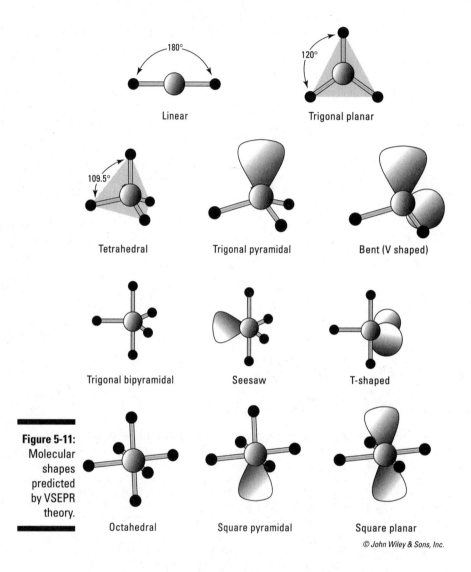

© John Wiley & Sons, Inc.

Figure 5-11: Molecular shapes predicted by VSEPR theory.

Consider one beautifully symmetrical shape predicted by VSEPR theory: the tetrahedron. Four equivalent pairs of electrons in the valence shell of an atom should distribute themselves into such a shape, with equal angles and an equal distance between each pair. But what sort of atom has four *equivalent* electron pairs in its valence shell? Aren't valence electrons distributed between different kinds of orbitals, like *s* and *p* orbitals? (We introduce these orbitals in Chapter 4.)

In order for VSEPR theory to make sense, it must be combined with another idea: hybridization. *Hybridization* refers to the mixing of atomic orbitals into new, hybrid orbitals of equal energy. Electron pairs occupy equivalent hybrid orbitals. It's important to realize that the hybrid orbitals are all equivalent, because that helps you understand the shapes that emerge from the electron pairs trying to distance themselves from one another. If electrons in a pure *p* orbital are trying to distance themselves from electrons in another *p* orbital *and* from electrons in an *s* orbital, the resulting shape may not be symmetrical, because *s* orbitals are different from *p* orbitals. But if all these electrons occupy identical hybrid orbitals (each orbital is a little bit *s* and a little bit *p*), then the resulting shape is more likely to be symmetrical.

Real molecules have all sorts of symmetrical shapes that just don't make sense if electrons truly occupy only "pure" orbitals (like *s* and *p*). The mixing of pure orbitals into hybrids allows chemists to explain the symmetrical shapes of real molecules with VSEPR theory. This kind of mixing must in some sense actually occur, as the case of methane, CH_4, makes clear.

The shape of methane, confirmed by experiment, is tetrahedral. The four C–H bonds of methane are of equal strength and are equidistant from each other. Now compare that description with the electron configuration of carbon in Figure 5-12. Carbon contains a filled 1*s* orbital, but this is an inner-shell orbital, so it doesn't impact the geometry of bonding. However, the valence shell of carbon contains one filled 2*s* orbital, two half-filled 2*p* orbitals, and one empty *p* orbital. Not the picture of equality. This configuration is inconsistent with the tetrahedral bonding geometry of carbon, meaning that the valence orbitals must hybridize.

Figure 5-12:
The electron
configuration
of carbon.

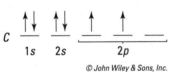

© John Wiley & Sons, Inc.

How can hydrogen atoms form four identical bonds with carbon? Conceptually, two things happen to produce the four equivalent orbitals necessary for methane:

✔ One of carbon's 2*s* electrons is promoted (sent to a higher-energy orbital) to the empty 2*p* orbital. This promotion results in four half-filled valence orbitals; the 2*s* orbital and three 2*p* orbitals now each contain one electron.

✔ The single 2*s* orbital combines with the three 2*p* orbitals to create four identical sp^3 hybrid orbitals. The fact that each sp^3 orbital is identical is important because VSEPR theory can now explain the symmetrical shape of methane: the tetrahedron.

So the shapes of real molecules emerge from the geometry of *valence orbitals* — the orbitals that bond to other atoms. Here's how to predict this molecular geometry:

1. **Count the number of lone pairs and bonding partners an atom actually has within a molecule. You can do this by looking at the Lewis structure.**

   In formaldehyde (CH_2O), for example, carbon bonds with two hydrogen atoms and double-bonds with one oxygen atom. So carbon effectively has three bonds present. Remember, the double bond counts as only one bond for the purposes of molecular geometry.

2. Look at Figure 5-13 and determine the molecular geometry of the compound based on the number of bonds and lone pairs.

You've determined that formaldehyde has three bonding pairs and no lone pairs of electrons present. Figure 5-13 indicates that the molecular geometry will be trigonal planar.

Number of Bonded Pairs on the Central Atom (Double and Triple Bonds Count Only as 1 Pair)	Number of Lone Electron Pairs on the Central Atom	Molecular Shape (Molecular Geometry)
1	Any amount	 Linear
2	0	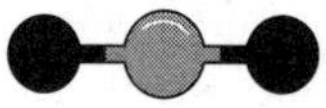 Linear
2	1	 Bent
2	2	 Bent
3	0	 Trigonal planar
3	1	 Trigonal pyramidal
4	0	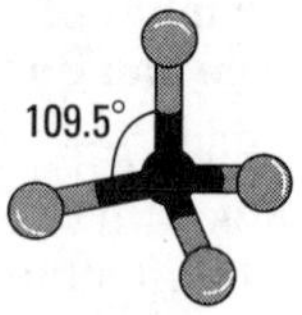 109.5° Tetrahedral

Figure 5-13: Use the number of bonded and lone pairs on the central atom to determine the molecular geometry.

© John Wiley & Sons, Inc.

EXAMPLE

Q. Methane, CH_4, has four hydrogen atoms bonded to a central carbon atom. Ammonia, NH_3, has three hydrogen atoms bonded to a central nitrogen atom. Using VSEPR theory, predict the molecular geometry of each compound.

A. After drawing out the Lewis structure of each molecule (though hopefully you can tell just from looking at the formulas), you see that CH_4 has four bonded pairs and zero lone pairs of electrons. This results in a **tetrahedral geometry.** NH_3 has three bonded pairs of electrons and one lone pair of electrons around the central nitrogen, resulting in a trigonal **pyramidal geometry.**

---

12. What's the hybridization of carbon in the following molecules?

a. carbon dioxide (CO_2)

b. formaldehyde (CH_2O)

c. methyl bromide (H_3CBr)

Solve It

13. Use Lewis structures and VSEPR theory to predict the geometry of the following molecules.

a. water (H_2O)

b. ethyne (C_2H_2)

c. carbon tetrachloride (CCl_4)

Solve It

Answers to Questions on Bonds

See how well you've bonded to the concepts in this chapter. If your answers don't overlap with the ones provided here, take another orbit through the questions.

1 Potassium (K) transfers its single valence electron to fluorine (F), yielding an ionic bond between K^+ and F^-, as in the following figure:

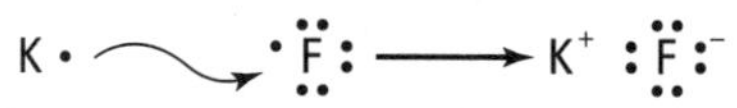

© John Wiley & Sons, Inc.

2 **Two lithium atoms each transfer a single electron to one sulfur atom** to yield the ionic compound Li_2S. As an alkali metal (Group IA), lithium easily gives up its single valence electron. As a Group VIA nonmetal, sulfur readily accepts two additional electrons into its valence shell.

3 As a salt, magnesium chloride ($MgCl_2$) is certainly an ionic compound. You can tell this easily because magnesium is a metal and chlorine is a nonmetal. Therefore, $MgCl_2$ dissolves to a greater extent in more polar solvents. Dissolved ions act as *electrolytes,* conducting electricity in solutions. The more brightly glowing bulb in the circuit containing the aqueous (water-based) solution suggests that more electrolytes are present in that solution. More salt dissolves in water than in rubbing alcohol because water is the more polar solvent.

4 Dichlorine is Cl_2, a compound formed when one chlorine atom bonds to another. Because each atom in the compound is of the same element, the two atoms have the same electronegativity. The difference in electronegativity between the two atoms is zero, so the bond between the two chlorine atoms must be covalent. Another easy way to tell that the bond is covalent is to recognize that chlorine is a nonmetal and is bonded to itself and that two nonmetals bonded together usually form a covalent bond. The electron dot structure of dichlorine is in the following figure:

© John Wiley & Sons, Inc.

5 **A coordinate covalent bond** forms between the aluminum and the hydrating water molecules. Aluminum is a Group IIIA element, so the aluminum (III) cation (with a charge of +3) formally has no valence electrons. The oxygen of water has lone pairs. Therefore, water molecules most likely hydrate the cation by donating lone pairs to form coordinate covalent bonds. In this respect, you can call the water molecules *ligands* of the metal ion.

6 You can see resonance structures for benzene in the following figure. Adjacent carbons in the ring are held together with either single or double covalent bonds, depending on the resonance structure. So in the resonance hybrid structure, each carbon-carbon bond is identical and is like a one-and-a-half bond rather than a single or double bond.

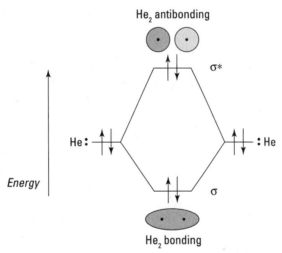

© John Wiley & Sons, Inc.

7 Take a look at the molecular orbital diagram for dihelium in the following figure. Each helium atom contributes two electrons to the molecular orbitals for a total of four electrons. Each molecular orbital holds two electrons, so both the low-energy bonding orbital and the high-energy antibonding orbitals are filled.

© John Wiley & Sons, Inc.

8 The total energy change to make dihelium from two separate helium atoms is the sum of the changes due to bonding and antibonding. Putting two electrons in the antibonding orbital costs more energy than is saved by putting two electrons in the bonding orbital. Therefore, going from two separate helium atoms to one molecule of dihelium requires an input of energy. Reactions spontaneously go to lower energy, not higher energy, so dihelium is unlikely to exist under normal conditions.

9 The reaction breaks one of the carbon-carbon bonds, replacing the double bond with two single bonds. The carbon-carbon bond that breaks is the pi bond, because the pi electrons are more accessible (above and below the axis of the bond) and because the pi bond is weaker than the sigma bond. Therefore, the weaker pi bond is replaced by a stronger sigma bond. In bonds, *weaker* means higher energy, and *stronger* means lower energy. The reaction moves from higher to lower energy, which is favorable.

10   The assignment of bond character depends on the difference in electronegativity between the atoms:

a. The difference in electronegativity between H and Cl is 0.9, so the bond is a **polar covalent bond.**

b. The difference in electronegativity between Ga and Ge is 0.2, so the bond is a **nonpolar covalent bond.**

c. The difference in electronegativity between O and O is 0.0, so the bond is a **nonpolar covalent bond.**

d. The difference in electronegativity between Na and Cl is 2.1, so the bond is an **ionic bond.**

e. The difference in electronegativity between C and O is 1.0, so the bond is a **polar covalent bond.**

11   Both molecules contain polar covalent bonds based on electronegativity difference between the atoms. The C–F bonds of CF_4 are slightly more polar than the H–O bonds of water due to the greater difference in electronegativity between carbon and flourine (1.5 versus 1.4), but the bonds are very close to being the same.

12   Here's the hybridization of carbon in the given molecules:

a. In CO_2, carbon is *sp* hybridized. CO_2 is linear, with the carbon double-bonded to each oxygen so that four electrons are constrained in each of two bond axes. As a reminder, even though two double bonds are present in carbon dioxide, each double bond counts as only one bonding domain.

b. In CH_2O, carbon is sp^2 hybridized, so the molecule is trigonal planar. The four electrons of the double bond are constrained within one bond axis, and two electrons each are within bonds to hydrogen.

c. In H_3CBr, carbon is sp^3 hybridized, so the molecule is shaped like a tetrahedron, similar to the shape of methane, CH_4. Each bond is separated from all the others by approximately 109°. Two electrons are constrained within each of four bonds.

13   The Lewis structures for the three molecules are in the following figure.

a. The oxygen of water is sp^3 hybridized, with two single bonds and two lone pairs, resulting in a **bent shape.**

b. Each carbon of ethyne is *sp* hybridized, with six electrons devoted to a triple bond and two devoted to a single bond with hydrogen, resulting in a **linear shape.**

c. Carbon tetrachloride is sp^3 hybridized, with all electron pairs involved in single bonds, resulting in a **tetrahedral shape.**

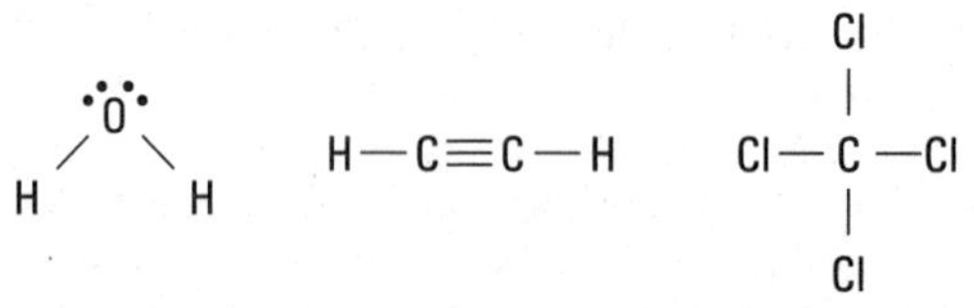

© John Wiley & Sons, Inc.

Chapter 6

Naming Compounds and Writing Formulas

Chemists give compounds very specific names. Sometimes these names seem overly specific. For example, what's the point of referring to "dihydrogen monoxide" when you can simply say "water"? First, some people simply believe this kind of thing sounds cool. Beyond dubious notions of coolness, however, lies a more important reason: Chemical names clue you in to chemical structures — but only if you know the code. Fortunately, the code, as you find out in this chapter, is pretty straightforward — no advanced cryptology required. (Which is also fortunate, because putting chemists and cryptologists in the same room could result in the kind of party you don't want to admit to having attended.)

Labeling Ionic Compounds and Writing Their Formulas

Chapter 5 discusses the way that *anions* (atoms with negative charge) and *cations* (atoms with positive charge) attract one another to form ionic bonds. Ionic bonds hold together ionic compounds. The anions and cations in a given ionic compound are important factors in how you name that compound.

Naming a simple ionic compound is easy. You pair the name of the cation with the name of the anion and then change the ending of the anion's name to *-ide*. The cation always precedes the anion in the final name. For example, the chemical name of NaCl (a compound made up of one sodium atom and one chlorine atom) is *sodium chloride*.

Of course, sodium chloride is more commonly known as table salt. Many compounds have such so-called *common names*. Common names aren't wrong, but they're less informative than chemical names. The name sodium chloride, properly decoded, tells you that you're dealing with a one-to-one ionic compound composed of sodium and chlorine. The name *table salt* just tells you one possible (albeit tasty) use for the compound.

Cations and anions combine in very predictable ways within ionic compounds, always acting to neutralize overall charge. Therefore, the name of an ionic compound implies more than just the identity of the atoms that make it up. It also helps you determine the correct chemical formula, which tells you the ratio in which the elements combine. Consider these two examples, both of which involve lithium:

- ✔ In the compound lithium fluoride (LiF), lithium and fluorine combine in a one-to-one ratio because lithium's +1 charge and fluorine's –1 charge cancel one another (neutralize) perfectly. By itself, the name *lithium fluoride* tells you only that the compound is made up of a lithium cation and a fluoride anion, but by comparing their charges, you can see that Li^+ and F^- neutralize each other in the one-to-one compound, LiF.

- ✔ If lithium combines with oxygen to form an ionic compound, then two lithium ions, each with +1 charge, are required to neutralize the –2 charge of the oxide anion. So the name lithium oxide implies the formula Li_2O, because a two-to-one ratio of lithium cation to oxide anion is necessary to produce a neutral compound.

Use the name of the ionic compound to identify the ions you're dealing with, and then combine those ions in the simplest way that results in a neutral compound.

Using a compound's name to identify the ions can be tricky when the cation is a metal. All Group B transition metals — with the exception of silver (which is always found as Ag^+) and zinc (always Zn^{2+}) — as well as several Group A elements on the right-hand side of the periodic table can take on a variety of charges. (Chapter 4 describes the periodic table.) Chemists use an additional naming device, the Roman numeral, to identify the charge state of the cation. A Roman numeral placed within parentheses after the name of the cation lists the positive charge of that cation. For example, copper (I) is copper with a +1 charge, and copper (II) is copper with a +2 charge. You may think this kind of distinction is simply chemical hair-splitting; who cares whether you're dealing with iron (II) bromide ($FeBr_2$) or iron (III) bromide ($FeBr_3$)? The difference matters because ionic compounds with different formulas (even those containing the same types of elements) can have very different properties.

Use Roman numerals only for Group B transition metals whose charge you don't know. Don't use Roman numerals if you know the charge of the metallic cation, including alkali metals such as lithium (Li^+) and sodium (Na^+) as well as alkaline earth metals such as calcium (Ca^{2+}) and magnesium (Mg^{2+}).

The trickiness of metals with variable charges also applies when you translate a chemical formula into the corresponding chemical name. Here are the two secrets to keep in mind:

- ✔ The name implies only the charge of the ions that make up an ionic compound.
- ✔ The charges of the ions determine the ratio in which they combine.

For example, you must do a little bit of sleuthing before assigning a name to the formula CrO. From periodic table trends (which we cover in Chapter 4), you know that oxygen brings a –2 charge to an ionic compound. Because O^{2-} combines with chromium in a one-to-one ratio within CrO, you know that chromium here must have the equal and opposite charge of +2. In this way, CrO can be electrically neutral (uncharged). So the chromium cation in CrO is Cr^{2+}, and the compound name is chromium (II) oxide. Simply calling the compound *chromium oxide* doesn't clearly identify the precise chromium cation in play. For example, chromium can assume a +3 charge. When Cr^{3+} combines with an oxide anion (which has a –2 charge), a different ion ratio is necessary to produce a neutral compound: Cr_2O_3, which is chromium (III) oxide. This equating of total positive and negative charges in an atom is called *balancing the charge*. After you've balanced the charges in the atom's formula, you can drop the charges on the individual ions.

EXAMPLE

Q. What is the name of Fe_2O_3?

A. **Iron (III) oxide.** The initial name is easy to come up with. Simply identify Fe as iron and change the ending of oxygen to *-ide* to get *iron oxide*. However, iron is a metal that has more than one possible charge, so you need a Roman numeral to indicate the charge of iron. Because you know that oxygen has a charge of –2 in ionic compounds and you have three oxygen atoms present, the total negative charge is –6. Two iron atoms are present, so each atom must have a charge of +3 for a total charge of +6 to balance out the oxygen's negative charge. To indicate this +3 charge, you add (III) after the iron.

Q. What is the formula for the compound tin (IV) fluoride?

A. **SnF_4.** The Roman numeral within parentheses tells you that you're dealing with Sn^{4+}. Because fluorine is a halogen, it always has a charge of –1 in ionic compounds, which means that four fluoride anions are necessary to cancel the four positive charges of a single tin cation. Therefore, the compound is SnF_4.

1. Name the following compounds:

a. MgF_2

b. LiBr

c. Cs_2O

d. CaS

Solve It

2. Name the following compounds that contain elements with variable charge. Don't forget to use Roman numerals!

a. FeF_2

b. CuBr

c. SnI_4

d. Mn_2O_3

Solve It

3. Translate the following names into chemical formulas:

a. Iron (III) oxide

b. Beryllium chloride

c. Tin (II) sulfide

d. Potassium iodide

Solve It

Getting a Grip on Ionic Compounds with Polyatomic Ions

A confession: Not all ions are of single atoms. To get a full education in chemical nomenclature, you must grapple with an irksome and all-too-common group of molecules called the *polyatomic ions,* which are made up of groups of atoms. Polyatomic ions, like single-element ions, tend to quickly combine with other ions to neutralize their charges. Unfortunately, you can't use any simple, periodic trend–type rules to figure out the charge of a polyatomic ion. You must — gulp — memorize them.

Table 6-1 summarizes the most common polyatomic ions, grouping them by charge.

Table 6-1	Common Polyatomic Ions		
–1 Charge	*–2 Charge*	*–3 Charge*	*+1 Charge*
Dihydrogen phosphate $(H_2PO_4^-)$	Hydrogen phosphate (HPO_4^{2-})	Phosphite (PO_3^{3-})	Ammonium (NH_4^+)
Acetate $(C_2H_3O_2^-)$	Oxalate $(C_2O_4^{2-})$	Phosphate (PO_4^{3-})	
Hydrogen sulfite (HSO_3^-)	Sulfite (SO_3^{2-})		
Hydrogen sulfate (HSO_4^-)	Sulfate (SO_4^{2-})		
Hydrogen carbonate (HCO_3^-)	Carbonate (CO_3^{2-})		
Nitrite (NO_2^-)	Chromate (CrO_4^{2-})		
Nitrate (NO_3^-)	Dichromate $(Cr_2O_7^{2-})$		

–1 Charge	–2 Charge	–3 Charge	+1 Charge
Cyanide (CN^-)	Silicate (SiO_3^{2-})		
Hydroxide (OH^-)			
Permanganate (MnO_4^-)			
Hypochlorite (ClO^-)			
Chlorite (ClO_2^-)			
Chlorate (ClO_3^-)			
Perchlorate (ClO_4^-)			

Notice in Table 6-1 that all the common polyatomic ions except ammonium have a negative charge ranging between –1 and –3. You also see a number of *-ite/-ate* pairs, such as chlorite and chlorate, phosphite and phosphate, and nitrite and nitrate. If you look closely at these pairs, you notice that the only difference between them is the number of oxygen atoms in each ion. Specifically, the *-ate* ion always has one more oxygen atom than the *-ite* ion but has the same overall charge.

To complicate your life further, polyatomic ions sometimes occur multiple times within the same ionic compound. How do you specify that your compound has two sulfate ions in a way that makes visual sense? Put the entire polyatomic ion formula in parentheses and then add a subscript outside the parentheses to indicate how many such ions you have, as in $\left(SO_4^{2-}\right)_2$.

When you write a chemical formula that involves polyatomic ions, you treat them just like other ions. You still need to balance charges to form a neutral atom. We're sorry to report that when you're converting from a formula to a name, you can't use any simple rule for naming polyatomic ions. You just have to memorize the entire table of polyatomic ions and their charges.

When you're tasked with writing the name of a formula containing a polyatomic ion, you follow all the same naming rules as listed previously except for one very simple change: You don't change the ending of any polyatomic ion. You leave it exactly as it written.

Q. Write the name of the formula $LiNO_3$.

A. **Lithium nitrate.** Because lithium is an alkali metal and the charge of alkali metals in Group I is always +1, you don't need to use Roman numerals to indicate the charge of lithium. You simply write *lithium* and then the name of the polyatomic ion, which is *nitrate*. You don't change the ending of the polyatomic ion name.

Q. Write the formula for the compound barium chlorite.

A. $Ba\left(ClO_2\right)_2$. Barium is an alkaline earth metal (Group IIA) and thus has a charge of +2. You should recognize *chlorite* as the name of a polyatomic ion. In fact, any anion name that doesn't end in *-ide* should scream *polyatomic ion* to you. As Table 6-1 shows, chlorite is ClO_2^-, which reveals that the chlorite ion has a –1 charge. Two chlorite ions are necessary to neutralize the +2 charge of a single barium cation, so the chemical formula is $Ba\left(ClO_2\right)_2$.

4. Name the following compounds that contain polyatomic ions:

a. $Mg_3(PO_4)_2$

b. $Pb(C_2H_3O_2)_2$

c. $Cr(NO_2)_3$

d. $(NH_4)_2C_2O_4$

e. $KMnO_4$

Solve It

5. Write the formula for the following compounds that contain polyatomic ions:

a. Potassium sulfate

b. Lead (II) dichromate

c. Ammonium chloride

d. Sodium hydroxide

e. Chromium (III) carbonate

Solve It

Naming Molecular (Covalent) Compounds and Writing Their Formulas

Nonmetals tend to form covalent bonds with one another (see Chapter 5 for details). Compounds made up of nonmetals held together by one or more covalent bonds are called *molecular* (or *covalent*) *compounds*.

Predicting how the atoms within molecules will bond with one another is a tricky endeavor because two nonmetals often can combine in multiple ratios. Carbon and oxygen, for example, can combine in a one-to-two ratio to form CO_2 (carbon dioxide), a harmless gas you emit every time you exhale. Alternatively, the same two elements can combine in a one-to-one ratio to form CO (carbon monoxide), a poisonous gas. Clearly, having names that distinguish between these (and other) molecular compounds is useful. The punishment for sloppy naming can be death. Or at least embarrassment.

Molecular compound names clearly specify how many of each type of atom participate in the compound. Table 6-2 lists the prefixes used to do so.

Table 6-2	Prefixes for Molecular Compounds		
Prefix	*Number of Atoms*	*Prefix*	*Number of Atoms*
mono-	1	hexa-	6
di-	2	hepta-	7
tri-	3	octa-	8
tetra-	4	nona-	9
penta-	5	deca-	10

You can attach the prefixes in Table 6-2 to any of the elements in a molecular compound, as exemplified by SO_3 (sulfur trioxide) and N_2O (dinitrogen monoxide). The second element in each compound receives the *-ide* suffix, as in ionic compounds (which we discuss earlier in this chapter). In the case of molecular compounds, where cations or anions aren't involved, the more electronegative element (in other words, the element that's closer to the upper right-hand corner of the periodic table) tends to be named second.

Note that the absence of a prefix from the first named element in a molecular compound implies that there's only one atom of that element. In other words, the prefix *mono-* is unnecessary for the first element only. You still have to attach a *mono-* prefix, when appropriate, to the names of subsequent elements.

Writing a formula for a molecular compound with a given name is much simpler than writing a formula for an ionic compound. In a molecular compound, the ratio in which the two elements combine is built into the name itself, and you don't need to worry about balancing charges. For example, the prefixes in the name *dihydrogen monoxide* imply that the chemical formula contains two hydrogens and one oxygen (H_2O).

Translating a formula into a name is equally simple. All you need to do is convert the subscripts into prefixes and attach them to the names of the elements that make up the compound. For example, for the compound N_2O_4, you simply attach the prefix *di-* to *nitrogen* to indicate the two nitrogen atoms and *tetra-* to *oxygen* to indicate the four oxygen atoms, giving you *dinitrogen tetroxide*.

Hydrogen is located on the far left of the periodic table, but it's actually a nonmetal. In keeping with this hydrogenic craziness, hydrogen can appear as either the first or second element in a *binary* (two-element) molecular compound, as shown by dihydrogen monosulfide (H_2S) and phosphorus trihydride (PH_3).

Q. What are the names of the compounds N_2O, SF_6, and Cl_2O_8?

A. **Dinitrogen monoxide, sulfur hexafluoride, and dichlorine octoxide.** Notice that none of these compounds contain any metals, which means that they're most certainly molecular compounds. The first compound contains two nitrogen atoms and one oxygen atom, so it's called *dinitrogen monoxide*. The second compound contains one sulfur and six fluorines. Because sulfur is the first named element, you don't need to include a *mono-* prefix. You simply name the compound *sulfur hexafluoride* (rather than *monosulfur hexafluoride*). Using the same methods, the third compound is named *dichlorine octoxide*.

Q. What is the formula of dicarbon tetrahydride?

A. C_2H_4. The prefixes in the name indicate the compound is molecular, so you don't need to worry about ionic charges. Just identify the element and the number of atoms based on the numerical prefix and then write it down. In this case, you have two carbons indicated by the *di-* and four hydrogens indicated by the *tetra-*.

6. Write the proper names for the following compounds:

 a. N_2H_4

 b. H_2S

 c. NO

 d. CBr_4

Solve It

7. Write the proper formulas for the following compounds:

 a. Silicon difluoride

 b. Nitrogen trifluoride

 c. Disulfur decafluoride

 d. Diphosphorus trichloride

Solve It

Addressing Acids

The chemical compounds known as *acids* have their own special naming system, but writing the names and formulas of acids really isn't that big of a deal. As long as you pay attention to the details, you won't have any problem naming acidic compounds. You more than likely already know the names of several very common acids, though perhaps you don't know their formulas. Table 6-3 lists several of the most common acids and their formulas. (Flip to Chapter 16 for full details on acids.)

Table 6-3	Common Acids
Name	**Formula**
Acetic acid	$HC_2H_3O_2$
Carbonic acid	H_2CO_3
Hydrochloric acid	HCl
Nitric acid	HNO_3
Phosphoric acid	H_3PO_4
Sulfuric acid	H_2SO_4

The two most common types of acids you encounter in a basic chemistry class are binary acids and oxy-acids:

✔ **Binary acids:** You can easily recognize a binary acid when you see hydrogen bonded to a nonmetallic element or polyatomic ion without oxygen present.

✔ **Oxy-acids:** Oxy-acids contain hydrogen bonded to a polyatomic ion containing oxygen.

To name a binary acid (no oxygen), use the following steps:

1. **Write the prefix *hydro-* at the beginning of the name.**

   Say you begin with the acid HCl. No oxygens are present, so the name starts with *hydro-*.

2. **Write the name of the anion.**

   If the name of the anion begins with a vowel, you drop the *o* in *hydro-* to avoid having two vowels next to each other. The anion in this case is chlorine, so you write *hydrochlorine*.

3. **Change the ending of the anion name to -*ic* and add *acid* to the end of the name.**

   So the name becomes *hydrochloric acid*.

The steps for naming an oxy-acid (in which oxygen is present) are a little different:

1. **Write the root name of the polyatomic anion.**

   Suppose you're naming HNO_3. Oxygen is present, so you don't begin by writing *hydro-*. Instead, simply write *nitrate*, because the anion is NO_3.

2. **If the polyatomic ion name ends in -*ate*, change the ending to -*ic*; if the polyatomic ion name ends in -*ite*, change the ending to -*ous*.**

   NO$_3$ is nitrate. It ends in -*ate*, so change the ending to -*ic*, giving you *nitric*.

3. **Put *acid* at the end of the name.**

   The name is *nitric acid*.

Q. What is the name of HBr?

A. **Hydrobromic acid.** HBr is a binary acid; it doesn't contain oxygen, so it isn't an oxy-acid. You begin by writing *hydro-*. Next, change the ending of the anion name, bromine, to -*ic* and write *bromic* after *hydro-*. Write *acid* at the end, and you're done!

Q. What is the name of H$_2$SO$_4$?

A. **Sulfuric acid.** H$_2$SO$_4$ is an oxy-acid because it contains oxygen. You begin by identifying SO$_4$ as sulfate. Sulfate ends in -*ate*, so you change the ending to -*ic*. You then add *acid* to the end of the name. (Don't worry about the fact that two hydrogen atoms are present at the beginning of the formula. Those two hydrogens are just in place to balance out the –2 charge of SO$_4$ with two positive charges. Each hydrogen ion has a charge of +1, for a total charge of +2.)

8. Write the proper names for the following binary acids:

   a. HF

   b. HI

   c. HCl

   d. HCN

   Solve It

9. Write the proper names for the following oxy-acids:

   a. H$_2$CO$_3$

   b. H$_2$SO$_3$

   c. HClO$_2$

   d. HNO$_3$

   Solve It

Mixing the Rules for Naming and Formula Writing

Have all the naming rules left you confused and frustrated? Do you feel like rebelling against the chemical conspiracy and liberating all chemicals from the oppressive confines of their formal names? All right, calm down. If you take a step back to look at the big picture, you see that the naming system is actually pretty logical and straightforward. It gives you a lot of valuable information when dealing with chemical compounds. The best way to really conquer the world of naming chemical compounds is to practice everything at once.

In short, you can think of this section as the big game. You've practiced and you've worked; you've put blood, sweat, and tears into these compounds. Now you're ready to show the world how it's done.

In reality, you aren't given nice subheadings telling you what type of compound a chemical is so that you know which set of naming rules to use. You need to figure it out. Here's a series of questions to help you write the names of all the wonderful types of chemicals in the previous sections:

1. **Does the formula begin with an *H*?** If so, use the rules presented earlier in "Addressing Acids." Be sure to identify whether the compound is a binary acid or an oxy-acid. If the compound doesn't begin with an *H*, move along to Question 2.

2. **Does the formula contain a metal (not hydrogen)?** If there's no metal, you're naming a molecular (covalent) compound, so you need to use the prefixes in Table 6-2. Be sure to change the ending of the second element to *-ide*. If there is a metal, then you're dealing with an ionic compound, so proceed to Question 3.

3. **Is the cation a transition metal (Group B) or a metal with a variable charge?** If the cation is a Group B metal (or other metal of variable charge, like tin), you need to use Roman numerals to specify its charge. See the earlier section "Labeling Ionic Compounds and Writing Their Formulas" for details. If the cation isn't a transition metal and you know the charge, you don't need to specify the charge with roman numerals.

4. **Is the anion a polyatomic ion?** If so, you have to recognize it as such and have its name memorized (or easily accessible in a nifty table such as Table 6-1). If the anion isn't a polyatomic ion, you use an *-ide* ending.

EXAMPLE

Q. Write the name of C_2H_7.

A. **Dicarbon heptahydride.** Carbon is a nonmetal, as is hydrogen, so this compound is covalent (molecular). That means you need to use prefixes when naming the formula. The compound has two carbons, so the carbon prefix is *di-*; it has seven hydrogens, so the hydrogen prefix is *hepta-*. You end the name by changing the ending of *hydrogen* to *-ide*.

Q. Write the formula of lead (IV) sulfate.

A. $Pb(SO_4)_2$. The cation lead is a metal with a variable charge, as indicated by the Roman numeral IV in parentheses. This classification means that lead has a charge of +4. SO_4 is the polyatomic ion sulfate with a charge of –2. To balance out the charges, you need two sulfate ions for each lead ion. Thus, the formula has one Pb ion with a total charge of +4 and two sulfate ions with a total charge of –4. To indicate the need for two polyatomic ions, you put parentheses around the sulfate ion and write the 2 as a subscript outside the parentheses.

10. Name each of the following compounds:

a. $PbCrO_4$

b. Mg_3P_2

c. $SrSiO_3$

d. H_2SO_4

e. Na_2S

f. B_3Se_2

g. HgF_2

h. $Ba_3(PO_4)_2$

Solve It

11. Translate the following names into chemical formulas:

a. Barium hydroxide

b. Tin (IV) bromide

c. Sodium sulfate

d. Phosphorus triiodide

e. Magnesium permanganate

f. Acetic acid

g. Nitrogen dihydride

h. Iron (II) chromate

Solve It

Beyond the Basics: Naming Organic Carbon Chains

One of the most common molecules studied in organic chemistry is the hydrocarbon. *Hydrocarbons* are compounds composed of carbon and hydrogen. The simplest of the hydrocarbons fall into the category of alkanes. *Alkanes* are chains of carbon molecules connected by single covalent bonds. Chapter 5 describes how single covalent bonds result when atoms share pairs of valence electrons. Because a carbon atom has four valence electrons, it's eager to donate those valence electrons to covalent bonds so it can receive four donated electrons in turn, filling carbon's valence shell. In other words, carbon really likes to form four bonds. In alkanes, each of these four is a single bond with a different partner.

As the name *hydrocarbon* suggests, these partners may be hydrogen or carbon. The simplest of the alkanes, called *continuous* or *straight-chain alkanes,* consist of one straight chain of carbon atoms linked with single bonds. Hydrogen atoms fill all the remaining bonds. Other types of alkanes include closed circles and branched chains, but we discuss the straight-chain alkanes here because they make clear the basic strategy for naming hydrocarbons. From the standpoint of naming, the hydrogen atoms in a hydrocarbon are more or less "filler atoms." Alkanes' names are based on the largest number of consecutively bonded carbon atoms, so the name of a hydrocarbon tells you about that molecule's structure.

To name a straight-chain alkane, simply match the appropriate chemical prefix with the suffix *-ane.* The prefixes, which relate to the number of carbons in the continuous chain, are listed in Table 6-4.

Table 6-4		Carbon Prefixes	
Number of Carbons	*Prefix*	*Chemical Formula*	*Alkane*
1	meth-	CH_4	methane
2	eth-	C_2H_6	ethane
3	prop-	C_3H_8	propane
4	but-	C_4H_{10}	butane
5	pent-	C_5H_{12}	pentane
6	hex-	C_6H_{14}	hexane
7	hept-	C_7H_{16}	heptane
8	oct-	C_8H_{18}	octane
9	non-	C_9H_{20}	nonane
10	dec-	$C_{10}H_{22}$	decane

The naming method in Table 6-4 tells you how many carbons are in the chain. Because you know that each carbon has four bonds and because you're fiendishly clever, you can deduce the number of hydrogen atoms in the molecule as well. Consider the carbon structure of pentane, for example, shown in Figure 6-1.

Figure 6-1:
Pentane's
carbon
skeleton.

© John Wiley & Sons, Inc.

Only four carbon-carbon bonds are required to produce the five-carbon chain of pentane. This situation leaves many bonds open — two for each interior carbon and three for each of the *terminal* carbons (the ones on either end of the chain). These open bonds are satisfied by carbon-hydrogen bonds, thereby forming a hydrocarbon, as shown in Figure 6-2.

Figure 6-2:
Pentane's
hydro-
carbon
structure.

© John Wiley & Sons, Inc.

If you add up the hydrogen atoms in Figure 6-2, you get 12. Therefore, pentane contains 5 carbon atoms and 12 hydrogen atoms.

The more complicated the organic molecule, the more important it is that you draw the molecular structure so you can visualize the molecule. In the case of straight-chain alkanes, the simplest of all organic molecules, you can remember a convenient formula for calculating the number of hydrogen atoms in the alkane without actually drawing the chain:

$$\text{Number of hydrogen atoms} = (2 \times \text{Number of carbon atoms}) + 2$$

You can refer to the same molecule in a number of different ways. For example, you can refer to pentane by its name (ahem . . . *pentane*); by its molecular formula, C_5H_{12}; or by the complete structure in Figure 6-2. Clearly, these names include different levels of structural detail. A *condensed structural formula* is another naming method, one that straddles the divide between a molecular formula and a complete structure. For pentane, the condensed structural formula is $CH_3CH_2CH_2CH_2CH_3$. This kind of formula assumes that you understand how straight-chain alkanes are put together. Here's the lowdown:

✔ Carbons on the end of a chain are bonded to only one other carbon, so they have three additional bonds that are filled by hydrogen and are labeled as CH_3 in a condensed formula.

✔ Interior carbons are bonded to two neighboring carbons and have only two hydrogen bonds, so they're labeled as CH_2.

Your chemistry teacher will probably require you to draw structures of alkanes when given their names and require you to name alkanes when given their structures. If your teacher fails to make such requests, ask to see his credentials. You may be dealing with an impostor.

Q. What is the name of the following structure, and what is its molecular formula?

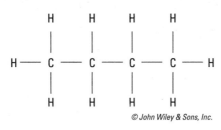

© John Wiley & Sons, Inc.

A. **Butane; C_4H_{10}.** First, count the number of carbons in the continuous chain. Four carbons are in the chain, and Table 6-4 helpfully points out that four-carbon chains earn the prefix *but-*. What's more, this molecule is an alkane (because it contains only single bonds), so it receives the suffix *-ane*. So what you've got is butane. With four carbon atoms in a straight chain, you need ten hydrogen atoms to satisfy all the carbon bonds, so the molecular formula of butane is C_4H_{10}.

12. What is the name of the following structure, and what is its molecular formula?

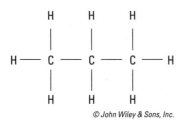

© John Wiley & Sons, Inc.

Solve It

13. Draw the structure of straight-chain octane.

Solve It

Answers to Questions on Naming Compounds and Writing Formulas

You survived all the practice questions on naming compounds. Now enjoy seeing how spot-on correct your answers were.

1 These compounds are all ordinary ionic compounds, so you simply need to pair the cation name with the anion name and change the anion name's ending to *-ide*.

a. **Magnesium fluoride**

b. **Lithium bromide**

c. **Cesium oxide**

d. **Calcium sulfide**

2 As the problem states, all the cations here are ones that can have varying amounts of positive charge, so you need to decipher their charges.

a. **Iron (II) fluoride.** The fluoride ion has a charge of –1. Because two fluorides are present here, the single iron ion must have a +2 charge.

b. **Copper (I) bromide.** The –1 charge of the bromide must be balanced by a +1 charge.

c. **Tin (IV) iodide.** The four iodide anions each have a –1 charge, so the tin cation must have a charge of +4.

d. **Manganese (III) oxide.** The three oxide anions here each have a charge of –2, giving an overall charge of –6. The manganese cations must carry a total of +6 charge, split between the two cations, so you must be dealing with Mn^{3+}.

3 The names translate into the following chemical formulas:

a. **Fe_2O_3.** Because the name specifies that you're dealing with Fe^{3+}, and because oxygen is always O^{2-}, you simply balance your charges to get Fe_2O_3.

b. **$BeCl_2$.** Beryllium is an alkaline earth metal with a charge of +2, while chlorine is a halogen with a charge of –1.

c. **SnS.** Because the name specifies that you're dealing with Sn^{2+}, and because sulfur is always S^{2-}, you simply balance your charges to get SnS.

d. **KI.** Potassium is an alkali metal with a charge of +1, and iodine is a halogen with a charge of –1.

4 Look up (or recall) the polyatomic ions in each compound, and specify the charge of the cation if it's a metal that can take on different charges.

a. **Magnesium phosphate.** Magnesium is the cation here, and the anion is PO_4. Table 6-1 tells you that this ion is the polyatomic ion phosphate (not to be confused with phosphite, PO_3).

b. **Lead (II) acetate.** Acetate has a –1 charge, and because two acetate ions are necessary to balance out the charge on the lead cation, you must be dealing with Pb^{2+}.

c. **Chromium (III) nitrite.** Nitrite has a –1 charge, and because three of them are necessary to balance out the charge on the chromium cation, you must be dealing with Cr^{3+}.

d. **Ammonium oxalate.** Here, both the cation and the anion are polyatomic ions — annoying but true.

e. **Potassium permaganate.** Potassium is the cation here, so all you need to do is look up the anion MnO_4 to find that it's called *permaganate*.

5 First, look up (or better, recall from memory) the charge of the polyatomic ion or ions, and then use subscripts as necessary to balance charges.

 a. K_2SO_4. Potassium has a charge of +1, and sulfate has a charge of –2. To balance these charges, you must double the number of potassiums. This balances the charges at +2 and –2.

 b. $PbCr_2O_7$. Here, the Roman numeral indicates that lead has a +2 charge. Dichromate is a polyatomic ion with a charge of –2, so you need only one of each to balance the charges.

 c. NH_4Cl. Ammonium is a +1 polyatomic ion, and chlorine is a –1 ion. These charges are balanced, so you need one of each.

 d. NaOH. Sodium has a charge of +1, and the hydroxide polyatomic ion has a charge of –1, so you need one of each.

 e. $Cr_2(CO_3)_3$. The Roman numeral indicates that chromium has a +3 charge. Carbonate is a polyatomic ion with a charge of –2. To balance this formula, you must come up with a multiple of 2 and 3. In this case, the least common multiple is 6, so you double the +3 and triple the –2. This results in two chromiums and three carbonates.

6 Using Table 6-2, translate the subscripts into prefixes. Omit the prefix *mono-* on the first named element in a compound where applicable.

 a. Dinitrogen tetrahydride

 b. Dihydrogen monosulfide

 c. Nitrogen monoxide

 d. Carbon tetrabromide

7 Using Table 6-2, translate the prefixes into subscripts. If the first named element lacks a prefix, assume that only one such atom exists per molecule.

 a. SiF_2

 b. NF_3

 c. S_2F_{10}

 d. P_2Cl_3

8 Identify the anion for each binary acid. Add *hydro-* to the beginning of it and change the ending of it to *-ic* (if necessary, drop the *o* in *hydro-* to avoid having two vowels next to each other). Then write *acid* at the end.

 a. Hydrofluoric acid

 b. Hydroiodic acid

 c. Hydrochloric acid

 d. Hydrocyanic acid

9 Identify the anion for each oxy-acid from the polyatomic ion chart in Table 6-1. If the polyatomic ion ends in *-ate,* change the ending to *-ic*. If the polyatomic ion ends in *-ite,* change the ending to *-ous*. Write *acid* at the end of the name. (Don't begin with *hydro-!* These aren't binary acids.)

 a. Carbonic acid

 b. Sulfurous acid

 c. Chlorous acid

 d. Nitric acid

10 Use the steps presented in "Mixing the Rules for Naming and Formula Writing" to guide yourself to a name from a formula.

a. Lead (II) chromate. Lead is a Group B element, and CrO_4^{2-} is a polyatomic ion. Therefore, you need to determine the charge on the lead and specify that charge with a Roman numeral. If you haven't memorized the name of CrO_4^{2-}, find it in Table 6-1. Because chromate combines with lead in a one-to-one ratio, you know that the charge on the lead must be +2.

b. Magnesium phosphide. This compound is a simple ionic compound because Mg is a non–Group B metal and because P isn't a polyatomic ion.

c. Strontium silicate. Sr is neither a Group B element nor a nonmetal, but SiO_3 forms a polyatomic ion.

d. Sulfuric acid. The H at the beginning of this formula is your clue that this compound is an acid. The H is then followed by the polyatomic ion sulfate. Because sulfate (SO_4^{-2}) ends in *-ate,* you change the ending to *-ic,* adding *ur* to make the name sound better. You then write *acid* at the end.

e. Sodium sulfide. Simply name the cation and change the ending of the (polyatomic) anion to *-ide*.

f. Triboron diselenide. Both boron and selenium are nonmetals, so this compound is a molecular compound. Therefore, you name it by using prefixes.

g. Mercury (II) fluoride. Here you have an ionic compound with a Group B metal, which means you need to use a Roman numeral. The charge on the mercury atom must be +2 because it combines with two fluoride anions, each of which must have a –1 charge.

h. Barium phosphate. This compound is an ionic compound that contains a polyatomic ion, phosphate.

11 Reverse your naming rules to deduce the chemical formula of each compound.

a. $Ba(OH)_2$. Barium is an ordinary metal, and hydroxide is a polyatomic ion with a charge of –1.

b. $SnBr_4$. The name indicates that you're dealing with Sn^{4+}. The bromide ion has a charge of –1, so you need four of them to balance the charge of a single tin cation.

c. Na_2SO_4. Sodium is a simple alkali metal, and sulfate is a polyatomic ion with a charge of –2.

d. PI_3. The prefixes indicate that this compound is a molecular compound containing a single phosphorus atom and three iodine atoms.

e. $Mg(MnO_4)_2$. Magnesium is a metal, and permanganate is a polyatomic ion with a charge of –1.

f. $HC_2H_3O_2$. Because this compound is an acid, you know the formula starts with an H. *Acetic* is nowhere on the periodic table, so you must assume it's a polyatomic ion. To figure out which one, change the *-ic* ending back to *-ate* to get *acetate.* Acetate has a charge of –1, and hydrogen has a charge of +1, so you need one of each to balance the charges.

g. NH_2. The *di-* prefix and the fact that both elements are nonmetals indicate that this compound is a molecular compound. Nitrogen, the first named element, lacks a prefix, so there must be only one nitrogen per molecule. The *di-* prefix indicates two hydrogen atoms per molecule.

h. $FeCrO_4$. The name tells you that you're dealing with Fe^{2+} and the polyatomic ion chromate, which has a charge of –2. A one-to-one ratio is sufficient to neutralize overall charge.

12   **The structure is propane; its molecular formula is C_3H_8.** The figure shows a three-carbon chain with only single bonds. Therefore, it's propane. Its molecular formula is C_3H_8.

13   The *oct-* prefix here tells you that this alkane is eight carbons long. Draw eight linked carbons and fill in the empty bonds with hydrogen. Your structure should look like one of these:

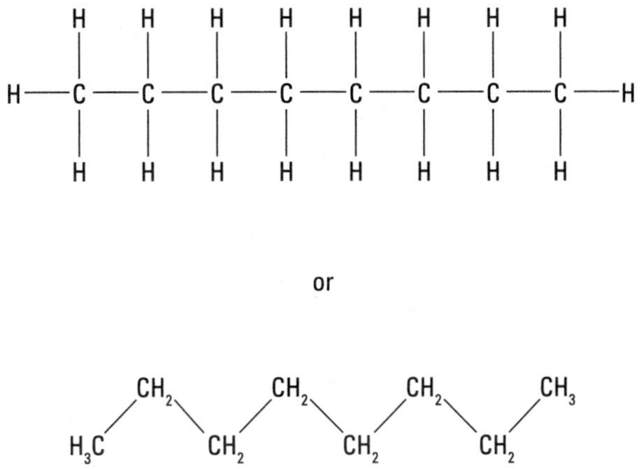

or

© John Wiley & Sons, Inc.

Chapter 7

Understanding the Many Uses of the Mole

· ·

In This Chapter

▶ Making particle numbers manageable with Avogadro's number

▶ Converting between masses, mole counts, and volumes

▶ Dissecting compounds with percent composition

▶ Moving from percent composition to empirical and molecular formulas

· ·

Chemists routinely deal with hunks of material containing trillions of trillions of atoms, but ridiculously large numbers can induce migraines. For this reason, chemists count particles (like atoms and molecules) in multiples of a quantity called the *mole*.

The mole is, without a doubt, the largest number you'll ever deal with on a regular basis. It represents 6.02×10^{23} particles. Chemists came up with this number to help work with the incredibly large number of atoms they deal with on a daily basis. Saying 2 moles of sodium chloride is far easier than saying 1.204×10^{24} atoms of sodium chloride.

The best way to understand the concept of the mole is to see it like any other representative quantity, such as a pair, a dozen, or a gross. For example, you know that 1 dozen of something is 12 of that same thing. Well, 1 mole of something is 6.02×10^{23} of that same thing. No different from a dozen — the mole is just much, much bigger. In this chapter, we explain what you need to know about moles.

For many mole conversions, you need to look up atomic masses on the periodic table (see Chapter 4). The atomic masses you see in different periodic tables may vary slightly, so for consistency, we've rounded all atomic mass values to two decimal places before plugging them into the equations. We round answers according to significant figure rules (see Chapter 1 for details).

The Mole Conversion Factor: Avogadro's Number

If 6.02×10^{23} strikes you as an unfathomably large number, then you're thinking about it correctly. It's larger, in fact, than the number of stars in the sky or the number of fish in the sea, and it's many, many times more than the number of people who've been born throughout all of human history. When you think about the number of particles in something as simple as, say, a cup of water, all your previous conceptions of "big numbers" are blown out of the water, as it were.

The number 6.02×10^{23}, known as *Avogadro's number,* is named after the 19th-century Italian scientist Amedeo Avogadro. Posthumously, Avogadro really pulled one off in giving his name to this number, because he never actually thought of it. The real brain behind Avogadro's number was that of a French scientist named Jean Baptiste Perrin. Nearly 100 years after Avogadro had his final pasta, Perrin named the number after Avogadro as an homage. Ironically, this humble act of tribute has misdirected the resentment of countless hordes of high school chemistry students to Avogadro instead of Perrin.

Avogadro's number is the conversion factor used to move between particle counts and numbers of moles. Notice that mole is abbreviated as *mol.*

$$\frac{1 \text{ mol}}{6.02 \times 10^{23} \text{ particles}}$$

The most common particles that you use with the mole are atoms and molecules. When encountering a problem that deals with a specific unit, like molecules, you just replace *particles* with the correct unit, as in the following example. Like all conversion factors, you can invert this fraction to move in the other direction, from moles to particles. (Flip to Chapter 2 for an introduction to conversion factors.)

Q. How many water molecules are in 1 tablespoon of water if the tablespoon holds 0.82 mol?

A. 4.9×10^{23} **molecules.** To convert from moles to particles, use a conversion factor with moles in the denominator and particles in the numerator:

$$(0.82 \text{ mol H}_2\text{O}) \left(\frac{6.02 \times 10^{23} \text{ molecules H}_2\text{O}}{1 \text{ mol H}_2\text{O}} \right) = 4.9 \times 10^{23} \text{ molecules H}_2\text{O}$$

The units of *moles* in the numerator of the first term and the denominator of the second term cancel, so you're left with the final answer of 4.9×10^{23} molecules. (If you need tips on multiplying in scientific notation, check out Chapter 1.)

---

1. If you have 1.3 mol of sodium (Na), how many atoms of sodium are present?

Solve It

2. If you have 7.9×10^{24} molecules of methane (CH_4), how many moles of methane are present?

Solve It

Doing Mass and Volume Mole Conversions

Chemists always begin a discussion about moles with Avogadro's number. They do this for two reasons. First, it makes sense to start the discussion with the way the mole was originally defined. Second, Avogadro's number is sufficiently large to intimidate the unworthy.

Still, for all its importance and intimidating size, Avogadro's number quickly grows tedious in everyday use. More interesting is the fact that 1 mole of an individual element turns out to possess exactly its atomic mass's worth of grams (see Chapter 3 for more about atomic mass). In other words, 1 mole of lithium atoms has a mass of about 6.94 grams, and 1 mole of helium atoms has a mass of about 4.00 grams. The same is true no matter where you wander through the corridors of the periodic table. The number listed as the atomic mass of an element also equals that element's *molar mass.* So in short, 1 mole of an individual element is equal to its atomic mass in grams.

Of course, chemistry involves the making and breaking of bonds (as you find out in Chapter 5), so talk of individual atoms only goes so far. How lucky, then, that calculating the molecular mass of a complex molecule is essentially no different from finding the mass per mole of individual elements. For example, a molecule of glucose ($C_6H_{12}O_6$) is assembled from 6 carbon atoms, 12 hydrogen atoms, and 6 oxygen atoms. To calculate the number of grams per mole of a complex molecule (such as glucose), simply do the following:

1. **Multiply the number of atoms per mole of the first element by its atomic mass.**

   In glucose, the first element is carbon. You'd multiply its atomic mass, 12.01, by the number of atoms, 6, for a total of 72.06.

2. **Multiply the number of atoms per mole of the second element by its atomic mass. Keep going until you've covered all the elements in the molecule.**

   Multiply hydrogen's atomic mass of 1.01 by the number of hydrogen atoms, 12, for a total of 12.12.

   The third element in glucose is oxygen, so you multiply 16.00, the atomic mass, by 6, the number of oxygen atoms, for a total of 96.00.

3. **Finally, add the masses together. The units that you use for molar mass is grams/mol.**

   Here are all the calculations for this example:

   $$\left(12.01 \, \frac{g}{mol}\right)(6 \text{ atoms carbon}) = 72.06 \text{ g/mol C}$$

   $$\left(1.01 \, \frac{g}{mol}\right)(12 \text{ atoms hydrogen}) = 12.12 \text{ g/mol H}$$

   $$\left(16.00 \, \frac{g}{mol}\right)(6 \text{ atoms oxygen}) = 96.00 \text{ g/mol O}$$

   $$72.06 \text{ g/mol C} + 12.12 \text{ g/mol H} + 96.00 \text{ g/mol O} = 180.18 \text{ g/mol } C_6H_{12}O_6$$

This kind of quantity, called the *gram molecular mass,* is exceptionally convenient for chemists, who are much more inclined to measure the mass of a substance than to count all the individual particles that make up a sample.

Chemists may distinguish between the molar masses of pure elements, molecular compounds, and ionic compounds by referring to them as the *gram atomic mass, gram molecular mass,* and *gram formula mass,* respectively. Don't be fooled! The basic concept behind each term is the same: molar mass.

It's all very good to find the mass of a solid or liquid and then go about calculating the number of moles in that sample. But what about gases? Let's not engage in phase discrimination; gases are made of matter, too, and their moles have the right to stand and be counted. Fortunately, there's a convenient way to convert between the moles of gaseous particles and their volume. Unlike gram atomic/molecular/formula masses, this conversion factor is constant — no matter what kinds of molecules make up the gas. Every gas has a volume of 22.4 liters per mole, regardless of the size of the gaseous molecules.

Before you start your hooray-chemistry-is-finally-getting-simple dance, understand that certain conditions apply to this conversion factor. For example, it's true only at *standard temperature and pressure* (STP), or 0°C and 1 atmosphere. Also, the figure of 22.4 L/mol applies only to the extent that a gas resembles an *ideal gas,* one whose particles have zero volume and neither attract nor repel one another. Ultimately, no gas is truly ideal, but many are so close to being so that the 22.4 L/mol conversion is very useful.

What if you want to convert between the volume of a gaseous substance and its mass or between the mass of a substance and the number of particles it contains? You already have all the information you need! To make these kinds of conversions, simply build a chain of conversion factors, converting units step by step from the ones you have (say, liters) to the ones you want (say, grams). You'll find that your chain of conversion factors always includes central links featuring units of moles. You can think of the mole as a family member who passes on what you've said, loudly barking into the ear of your nearly deaf grandmother because you have laryngitis and can't speak any louder. Without such a central translator, your message would no doubt be misinterpreted. "Grandma, how was your day?" would be received as "Grandma, you want to eat clay?" So unless you're bent on force-feeding clay to your grandmother, do not attempt to convert directly from volume to mass, from mass to particles, or any other such shortcut. Use your translator, the mole.

Q. Convert 65 g of carbon dioxide (CO_2) to moles of carbon dioxide.

A. **1.5 mol CO_2.** Before beginning this problem, you need to determine the molecular mass of CO_2. Do this by adding the atomic mass of 1 carbon atom ($1 \times 12.01 = 12.01$) and the atomic mass of 2 oxygens ($2 \times 16.00 = 32.00$) for a total of 44.01 g/mol. Then use this molecular mass to create a conversion factor and determine the moles. Be sure your units cancel out to ensure your conversion factor is set up correctly.

$$(65 \text{ g } CO_2) \left(\frac{1 \text{ mol } CO_2}{44.01 \text{ g } CO_2} \right) = 1.5 \text{ mol } CO_2$$

Q. Convert 26.7 g of carbon dioxide (CO_2) to liters of carbon dioxide gas.

A. **13.6 L CO_2.** The first thing you should notice about this problem is that it requires two steps. You must first convert from grams of CO_2 to moles of CO_2 using the molecular mass of carbon dioxide. Find this molecular mass by adding the atomic mass of 1 carbon atom ($1 \times 12.01 = 12.01$) and the atomic mass of 2 oxygens ($2 \times 16.00 = 32.00$) for a total of 44.01 g/mol. After converting to moles, you convert to liters of CO_2 gas using the molar volume constant, 22.4 L/mol. However, instead of doing each of these steps in separate calculations, you can combine them so you do everything on one line. As always, be sure your units cancel out so that you end up with your goal, in this case liters of carbon dioxide.

$$(26.7 \text{ g } CO_2) \left(\frac{1 \text{ mol } CO_2}{44.01 \text{ g } CO_2} \right) \left(\frac{22.4 \text{ L } CO_2}{1 \text{ mol } CO_2} \right)$$
$$= 13.6 \text{ L } CO_2$$

3. Do the conversion:

    a. How many grams are present in 2.6 mol of lithium?

    b. How many moles are present in 85.2 g of sodium chloride (NaCl)?

Solve It

4. Do the conversion:

    a. How many liters of gas are present in 4.3 mol of oxygen gas (O_2)?

    b. How many moles of gas are present in 64.3 L of nitrogen gas (N_2)?

Solve It

5. Do the conversion:

    a. How many liters of hydrogen gas (H_2) are present in 76.2 g of hydrogen gas?

    b. How many atoms are present in 10 g of lithium?

Solve It

Determining Percent Composition

Chemists are often concerned with precisely what percentage of a compound's mass consists of one particular element. Lying awake at night, uttering prayers to Avogadro, they fret over this quantity, called *percent composition*. Calculating percent composition is trickier than you may think. Consider the following problem, for example.

The human body is composed of 60 to 70 percent water, and water contains twice as many hydrogen atoms as oxygen atoms. If two-thirds of every water molecule is hydrogen and if water makes up 60 percent of the body, it seems logical to conclude that hydrogen makes up 40 percent of the body. Yet hydrogen is only the third most abundant element in the body by mass. What gives?

Oxygen is 16 times more massive than hydrogen, so equating *atoms* of hydrogen and *atoms* of oxygen is a bit like equating a toddler to a sumo wrestler. When the doors of the elevator won't close, the sumo wrestler is the first one you should kick out, weep though he may.

Within a compound, it's important to sort out the atomic toddlers from the atomic sumo wrestlers. To do so, follow three simple steps:

1. **Calculate the molar mass of the compound, as we explain in the preceding section.**

   This value represents the total mass of the compound.

2. **Determine the total mass of each element in the compound, being sure to account for all atoms present.**

   For example, if you have three carbon atoms, be sure to multiply the mass of carbon (12.01 g/mol) by 3. These values represent the individual mass of each element in the compound.

3. **Use the following formula to find the percent composition for each element in the compound:**

$$\% \text{ composition of element} = \left(\frac{\text{Total mass of element}}{\text{Total mass of entire compound}} \right) \times 100$$

Q. Calculate the percent composition for each element in sodium sulfate, Na_2SO_4.

A. **Na: 32.4%, S: 22.6%, O: 45.0%.** To begin, determine the molecular mass of sodium sulfate. Do this by adding the masses of 2 sodium atoms plus 1 sulfur and 4 oxygens together:

$$\left(22.99 \frac{g}{mol} \right) (2 \text{ atoms Na}) = 45.98 \text{ g/mol Na}$$

$$\left(32.06 \frac{g}{mol} \right) (1 \text{ atoms S}) = 32.06 \text{ g/mol S}$$

$$\left(16.00 \frac{g}{mol} \right) (4 \text{ atoms O}) = 64.00 \text{ g/mol O}$$

$$45.98 \text{ g/mol Na} + 32.06 \text{ g/mol S} + 64.00 \text{ g/mol O} = 142.04 \text{ g/mol } Na_2SO_4$$

You've now also determined the total mass of each element present in the formula. Now all you need to do is plug each element's value into the percent composition formula and solve:

$$\% \text{ composition of Na} = \left(\frac{45.98 \text{ g Na}}{142.04 \text{ g } Na_2SO_4} \right) \times 100 = 32.4\% \text{ Na}$$

$$\% \text{ composition of S} = \left(\frac{32.06 \text{ g S}}{142.04 \text{ g } Na_2SO_4} \right) \times 100 = 22.6\% \text{ S}$$

$$\% \text{ composition of O} = \left(\frac{64.00 \text{ g O}}{142.04 \text{ g } Na_2SO_4} \right) \times 100 = 45.0\% \text{ Na}$$

6. Calculate the percent composition of potassium chromate, K_2CrO_4.

Solve It

Calculating Empirical Formulas

What if you don't know the formula of a compound? Chemists sometimes find themselves in this disconcerting scenario. Instead of cursing Avogadro (or perhaps *after* doing so), they analyze samples of the frustrating unknown to identify the percent composition. From there, they calculate the ratios of different types of atoms in the compound. They express these ratios as an *empirical formula*, the lowest whole-number ratio of elements in a compound.

Here's how to find an empirical formula when given percent composition:

1. **Assume that you have 100 g of the unknown compound.**

   The beauty of this little trick is that you conveniently gift yourself with the same number of grams of each elemental component as its contribution to the percent composition. For example, if you assume that you have 100 g of a compound composed of 60.3% magnesium and 39.7% oxygen, you know that you have 60.3 g of magnesium and 39.7 g of oxygen. (The only time you don't do this is if the problem specifically gives you the masses of each element present in the unknown compound.)

2. **Convert the masses from Step 1 into moles using the molar mass.**

   See the earlier section "Doing Mass and Volume Mole Conversions."

3. **Determine which element has the smallest mole value. Then divide all the mole values you calculated in Step 2 by this smallest value.**

   This division yields the mole ratios of the elements of the compound.

4. **If any of your mole ratios aren't whole numbers, multiply all numbers by the smallest possible factor that produces whole-number mole ratios for all the elements.**

   For example, if you have 1 nitrogen atom for every 0.5 oxygen atoms in a compound, the empirical formula is not $N_1O_{0.5}$. Such a formula casually suggests that an oxygen atom has been split, something that would create a small-scale nuclear explosion. Though impressive sounding, this scenario is almost certainly false. Far more likely is that the atoms of nitrogen and oxygen are combining in a 1 : 0.5 *ratio* but do so in a larger but equivalent ratio of 2 : 1. The empirical formula is thus N_2O.

Because the original percent composition data is typically experimental, expect to see a bit of error in the numbers. For example, 2.03 is probably within experimental error of 2, 2.99 is probably 3, and so on.

5. **Write the empirical formula by attaching these whole-number mole ratios as subscripts to the chemical symbol of each element. Order the elements according to the general rules for naming ionic and molecular compounds.**

We describe the naming rules in Chapter 6.

Q. What is the empirical formula of a substance that is 40.0% carbon, 6.7% hydrogen, and 53.3% oxygen by mass?

A. **CH$_2$O.** For the sake of simplicity, assume that you have a total of 100 g of this mystery compound. Therefore, you have 40.0 g of carbon, 6.7 g of hydrogen, and 53.3 g of oxygen. Convert each of these masses to moles by using the gram atomic masses of C, H, and O:

$$(40.0 \text{ g C}) \left(\frac{1 \text{ mol C}}{12.01 \text{ g C}} \right) = 3.33 \text{ mol C}$$

$$(6.7 \text{ g H}) \left(\frac{1 \text{ mol H}}{1.01 \text{ g H}} \right) = 6.6 \text{ mol H}$$

$$(53.3 \text{ g O}) \left(\frac{1 \text{ mol O}}{16.00 \text{ g O}} \right) = 3.33 \text{ mol O}$$

Notice that the carbon and oxygen mole numbers are the same, so you know the ratio of these two elements is 1:1 within the compound. Next, divide all the mole numbers by the smallest among them, which is 3.33. This division yields

$$\frac{3.33 \text{ mol C}}{3.33} = 1 \text{ mol C}, \ \frac{6.6 \text{ mol H}}{3.33} = 2 \text{ mol H}, \ \frac{3.33 \text{ mol O}}{3.33} = 1 \text{ mol O}$$

The compound has the empirical formula CH$_2$O. The actual number of atoms within each particle of the compound is some multiple of the numbers expressed in this formula.

---

7. Calculate the empirical formula of a compound with a percent composition of 88.9% oxygen and 11.1% hydrogen.

Solve It

8. Calculate the empirical formula of a compound with a percent composition of 40.0% sulfur and 60.0% oxygen.

Solve It

Using Empirical Formulas to Find Molecular Formulas

Many compounds in nature, particularly compounds made of carbon, hydrogen, and oxygen, are composed of atoms that occur in numbers that are multiples of their empirical formula. In other words, their empirical formulas don't reflect the actual numbers of atoms within them; instead, they reflect only the ratios of those atoms. What a nuisance! Fortunately, this is an old nuisance, so chemists have devised a means to deal with it. To account for these annoying types of compounds, chemists are careful to differentiate between an empirical formula and a molecular formula. A *molecular formula* uses subscripts that report the actual number of each type of atom in a molecule of the compound (a *formula unit* accomplishes the same thing for ionic compounds).

Molecular formulas are associated with gram molecular masses that are simple whole-number multiples of the corresponding *empirical formula mass.* For example, a molecule with the empirical formula CH_2O has an empirical formula mass of about 30 g/mol (12 for the carbon + 2 for the two hydrogens + 16 for the oxygen). The molecule may have a molecular formula of CH_2O, $C_2H_4O_2$, $C_3H_6O_3$, or the like. As a result, the compound may have a gram molecular mass of 30 g/mol, 60 g/mol, 90 g/mol, or another multiple of 30 g/mol.

You can't calculate a molecular formula based on percent composition alone. If you attempt to do so, Avogadro and Perrin will rise from their graves, find you, and slap you 6.02×10^{23} times per cheek. You can clearly see the folly of such an approach by comparing formaldehyde with glucose. The two compounds have the same empirical formula, CH_2O, but different molecular formulas, CH_2O and $C_6H_{12}O_6$, respectively. Glucose is a simple sugar, the one made by photosynthesis and the one broken down during cellular respiration. You can dissolve it in your coffee with pleasant results. Formaldehyde is a carcinogenic component of smog. Solutions of formaldehyde have historically been used to embalm dead bodies. Dissolving formaldehyde in your coffee is not advised. In other words, molecular formulas differ from empirical formulas, and the difference is important in the real world.

To determine a molecular formula, you must know the gram formula mass of the compound as well as the empirical formula (or enough information to calculate it yourself from the percent composition; see the preceding section for details). With these tools in hand, calculating the molecular formula involves three steps:

1. **Calculate the empirical formula mass.**

2. **Divide the gram molecular mass by the empirical formula mass.**

3. **Multiply each of the subscripts within the empirical formula by the number calculated in Step 2.**

Q. What is the molecular formula of a compound that has a gram molecular mass of 34 g/mol and the empirical formula HO?

A. H_2O_2. The empirical formula mass is 17.01 g/mol. You determine this number by finding the mass of HO (1 hydrogen atom and 1 oxygen atom).

H: $1.01 \cdot 1 = 1.01$

O: $16.00 \cdot 1 = 16.00$

$1.01 + 16.00 = 17.01$ g/mol HO

Dividing the gram molecular mass by this value yields the following:

$$\frac{\text{Molecular formula mass}}{\text{Empirical formula mass}} = \frac{34 \frac{g}{mol}}{17.01 \frac{g}{mol}} \approx 2$$

Multiplying the subscripts within the empirical formula by this number gives you the molecular formula H_2O_2. This formula corresponds to the compound *hydrogen peroxide*.

9. What is the molecular formula of a compound that has a gram formula mass of 78 g/mol and the empirical formula NaO?

Solve It

10. A compound has a percent composition of 49.5% carbon, 5.2% hydrogen, 16.5% oxygen, and 28.8% nitrogen. The compound's gram molecular mass is 194.2 g/mol. What are the empirical and molecular formulas?

Solve It

Answers to Questions on Moles

Following are the answers to the practice problems in this chapter.

1 **7.8×10^{23} atoms Na.** To calculate the answer, make sure you use the correct conversion factor with Avogadro's number. The units of *moles* in the first term and in the denominator of the second term cancel, leaving you the answer in number of atoms:

$$(1.3 \text{ mol Na}) \left(\frac{6.02 \times 10^{23} \text{ atoms Na}}{1 \text{ mol Na}} \right) = 7.8 \times 10^{23} \text{ atoms Na}$$

2 **13 mol CH_4.** To calculate the answer, use the correct conversion factor with Avogadro's number. In this case, you flip it so that moles are on top. The units of molecules in the first term and in the denominator of the second term cancel, giving you the answer in moles:

$$\left(7.9 \times 10^{24} \text{ molecules CH}_4 \right) \left(\frac{1 \text{ mol CH}_4}{6.02 \times 10^{23} \text{ molecules CH}_4} \right) = 13 \text{ mol CH}_4$$

3 Complete the conversions:

a. 18 g Li. To solve this problem, you need to take your mole value of Li and multiply it by the molar mass of the element lithium (which is approximately 6.94, according to the periodic table). Ensure units cancel to make sure you set up your conversion factor correctly.

$$(2.6 \text{ mol Li}) \left(\frac{6.94 \text{ g Li}}{1 \text{ mol Li}} \right) = 18 \text{ g Li}$$

b. 1.46 mol NaCl. Determine the formula mass of NaCl. To do this, add the molar masses of 1 sodium (22.99 g) and 1 chlorine (35.45 g) together for a formulas mass of 54.88 g. Then divide the given initial mass by the formula mass of NaCl:

$$(85.2 \text{ g NaCl}) \left(\frac{1 \text{ mol NaCl}}{58.44 \text{ g NaCl}} \right) = 1.46 \text{ mol NaCl}$$

4 Complete the conversions:

a. 96 L O_2. To solve this problem, you need to take your mole value of O_2 and multiply it by the molar volume constant of 22.4 L/mol. Ensure units cancel to make sure you set up your conversion factor correctly.

$$(4.3 \text{ mol O}_2) \left(\frac{22.4 \text{ L O}_2}{1 \text{ mol O}_2} \right) = 96 \text{ L O}_2$$

b. 2.87 mol N_2. Perform the same conversion as in part (a), except this time flip the conversion factor so that *liters* is on the bottom:

$$(64.3 \text{ L N}_2) \left(\frac{1 \text{ mol N}_2}{22.4 \text{ L N}_2} \right) = 2.87 \text{ mol N}_2$$

5 Complete the conversions:

a. 845 L H₂. This is a two-step problem. The first step requires converting from grams to moles of hydrogen. Then convert from moles to liters of hydrogen. To determine the molecular mass of hydrogen (H_2), add together the masses of 2 hydrogen atoms (1.01) for a total of 2.02.

$$(76.2 \text{ g } H_2) \left(\frac{1 \text{ mol } H_2}{2.02 \text{ g } H_2} \right) \left(\frac{22.4 \text{ L } H_2}{1 \text{ mol } H_2} \right) = 845 \text{ L } H_2$$

b. 8.7×10^{23} atoms Li. First convert grams of lithium to moles of lithium by dividing the initial value by the molar mass of lithium off the periodic table. Then multiply by Avogadro's number to determine the number of particles (atoms in this case).

$$(10 \text{ g Li}) \left(\frac{1 \text{ mol Li}}{6.94 \text{ g Li}} \right) \left(\frac{6.02 \times 10^{23} \text{ atoms Li}}{1 \text{ mol Li}} \right) = 8.7 \times 10^{23} \text{ atoms Li}$$

6 **K: 40.3%, Cr: 26.8%, O: 33.0%.** First, calculate the gram molecular mass of potassium chromate, which comes to 194.2 g/mol:

$$\left(39.10 \frac{g}{mol} \right) (2 \text{ atoms K}) = 78.20 \text{ g/mol K}$$

$$\left(52.00 \frac{g}{mol} \right) (1 \text{ atoms Cr}) = 52.00 \text{ g/mol Cr}$$

$$\left(16.00 \frac{g}{mol} \right) (4 \text{ atoms O}) = 64.00 \text{ g/mol O}$$

$$78.20 \text{ g/mol K} + 52.00 \text{ g/mol Cr} + 64.00 \text{ g/mol O} = 194.20 \text{ g/mol } K_2CrO_4$$

In a 194.20 g sample, 78.20 g is potassium, 52.00 g is chromium, and 64.00 g is oxygen. Divide each of these masses by the gram molecular mass, and then multiply by 100 to get the percent composition:

$$\% \text{ composition of K} = \left(\frac{78.20 \text{ g Na}}{194.2 \text{ g } K_2CrO_4} \right) \times 100 = 40.3\% \text{ Na}$$

$$\% \text{ composition of Cr} = \left(\frac{52.00 \text{ g Cr}}{194.2 \text{ g } K_2CrO_4} \right) \times 100 = 26.8\% \text{ S}$$

$$\% \text{ composition of O} = \left(\frac{64.00 \text{ g O}}{194.2 \text{ g } K_2CrO_4} \right) \times 100 = 33.0\% \text{ Na}$$

Note: If you rounded your percentages properly, they add to 100.1%, not 100%. If you do away with rounding, you get exactly 100%. Rounding is common practice, though, so don't be too worried if your answer is off by a tenth or two.

7 **H_2O.** First, assume that you have 88.9 g of oxygen and 11.1 g of hydrogen in a 100 g sample. Then convert each of these masses into moles by using the gram atomic masses of oxygen and hydrogen:

$$(88.9 \text{ g O}) \left(\frac{1 \text{ mol O}}{16.00 \text{ g O}} \right) = 5.56 \text{ mol O}$$

$$(11.1 \text{ g H}) \left(\frac{1 \text{ mol H}}{1.01 \text{ g H}} \right) = 11.0 \text{ mol H}$$

Next, divide each of these mole quantities by the smallest among them, 5.56 mol:

$$\frac{5.56 \text{ mol O}}{5.56} = 1, \quad \frac{11.0 \text{ mol H}}{5.56} = 2$$

Attach these quotients as subscripts and list the atoms properly. This yields H_2O. The compound is water.

8 SO_3. Following the same procedure as in Question 7, you calculate 1.25 mol of sulfur and 3.75 mol of oxygen. You must change your percentages to mass and then divide by the molar mass of each element.

$$(40.0 \text{ g S}) \left(\frac{1 \text{ mol S}}{32.06 \text{ g S}} \right) = 1.25 \text{ mol S}$$

$$(60.0 \text{ g O}) \left(\frac{1 \text{ mol O}}{16.00 \text{ g O}} \right) = 3.75 \text{ mol O}$$

Dividing each of these quantities by 1.25 mol (the smallest quantity) yields $1.25/1.25 = 1$ sulfur and $3.75/1.25 = 3$ oxygen, or a mole ratio of 1 sulfur to 3 oxygens. The compound is SO_3, sulfur trioxide.

9 Na_2O_2. First, find the empirical formula mass of NaO, which is 38.99 g/mol. You determine this by adding one Na (22.99) to one O (16.00). Then divide the gram formula mass of the mystery compound, 78 g/mol, by this empirical formula mass to obtain the quotient, 2. Multiply each of the subscripts within the empirical formula by this number to obtain Na_2O_2. You've just found the molecular formula for sodium peroxide.

10 $C_4H_5N_2O$ is the empirical formula; $C_8H_{10}N_4O_2$ is the molecular formula. You're not directly given the empirical formula of this compound, but you *are* given the percent composition. Using the percent composition, you can calculate the empirical formula. To do so, assume that you have 100 g of the substance, giving you 49.5 g of carbon, 5.2 g of hydrogen, 16.5 g of oxygen, and 28.8 g of nitrogen. Then divide these masses by the atomic mass of each element, giving you the following numbers of moles:

$$\frac{49.5 \text{ g C}}{12.01 \text{ g/mol}} = 4.125 \text{ mol C}$$

$$\frac{5.2 \text{ g C}}{1.01 \text{ g/mol}} = 5.2 \text{ mol H}$$

$$\frac{16.5 \text{ g O}}{16.00 \text{ g/mol}} = 1.031 \text{ mol O}$$

$$\frac{28.8 \text{ g O}}{14.01 \text{ g/mol}} = 2.057 \text{ mol N}$$

Finally, divide each of these mole values by the lowest among them, 1.031. You get 4.0 mol carbon, 5.0 mol hydrogen, 1.0 mol oxygen, and 2.0 mol nitrogen, giving you the empirical formula $C_4H_5N_2O$.

The empirical formula mass is 97.1 g/mol, which you calculate by multiplying the number of atoms of each element in the compound by the element's atomic mass and adding them all up:

$$C: 12.01 \cdot 4 = 48.04$$
$$H: 1.01 \cdot 5 = 5.05$$
$$N: 14.01 \cdot 2 = 28.02$$
$$O: 16.00 \cdot 1 = 16.00$$

$$48.04 + 5.05 + 28.02 + 16 = 97.1 \text{ g/mol}$$

Dividing the gram molecular mass you were given (194.2 g/mol) by this empirical formula mass yields the quotient, 2. Multiplying each of the subscripts in the empirical formula by 2 produces the molecular formula, $C_8H_{10}N_4O_2$. The common name for this culturally important compound is caffeine.

Chapter 8

Getting a Grip on Chemical Equations

• •

In This Chapter
▶ Reading, writing, and balancing chemical equations
▶ Recognizing five types of reactions and predicting products
▶ Charging through net ionic equations

• •

Chapters 5, 6, and 7 focus on chemical compounds and the bonds that bind them. You can think of a compound in two different ways:

✔ As the product of one chemical reaction
✔ As a starting material in another chemical reaction

In the end, chemistry is about action — about the breaking and making of bonds. Chemists describe action by using *chemical equations,* sentences that say who reacted with whom and who remained when the smoke cleared. This chapter explains how to read, write, balance, and predict the products of these action-packed chemical sentences.

Translating Chemistry into Equations and Symbols

In general, all chemical equations are written in the basic form

$$\text{Reactants} \rightarrow \text{Products}$$

where the arrow in the middle means *yields.* The basic idea is that the reactants react and the reaction produces products. By *reacting,* we simply mean that bonds within the reactants are broken, to be replaced by new and different bonds within the products.

Chemists fill chemical equations with symbols because they think it looks cool and, more importantly, because the symbols pack a lot of meaning into a small space. Table 8-1 summarizes the most important symbols you find in chemical equations.

Table 8-1	Symbols Commonly Used in Chemical Equations
Symbol	*Explanation*
+	The plus sign separates two reactants or products.
→	The *yields* symbol separates the reactants from the products. The single arrowhead suggests the reaction occurs in only one direction.
↔	A two-way *yields* symbol means the reaction can occur reversibly, in both directions. You may also see this symbol written as two stacked arrows with opposing arrowheads.
(s)	A reactant or product followed by this symbol exists as a solid.
(l)	A reactant or product followed by this symbol exists as a liquid.
(g)	A reactant or product followed by this symbol exists as a gas.
(aq)	A reactant or product followed by this symbol exists in aqueous solution, dissolved in water.
Δ	This symbol, usually written above the *yields* symbol, signifies that heat is added to the reactants.
Ni, LiCl	Sometimes a chemical symbol (such as those for nickel or lithium chloride here) is written above the *yields* symbol. This means that the indicated chemical was added as a catalyst. Catalysts speed up reactions but do not otherwise participate in them.

After you understand how to interpret chemical symbols, the names of compounds (see Chapter 6), and the symbols in Table 8-1, you can understand almost anything. You're equipped, for example, to decode a chemical equation into an English sentence describing a reaction. Conversely, you can translate an English sentence into the chemical equation it describes. When you're fluent in this language, you regrettably won't be able to talk to the animals; you will, however, be able to describe their metabolism in great detail.

Q. Write out the chemical equation for the following sentence:

Solid iron (III) oxide reacts with gaseous carbon monoxide to produce solid iron and gaseous carbon dioxide.

A. $Fe_2O_3(s) + CO(g) \rightarrow Fe(s) + CO_2(g)$. First, convert each formula into the written name for the compound (check out Chapter 6 for details). Next, annotate the

physical state of the compound if it's provided. Then group the compounds into "reactant" and "product" categories. Things that react are reactants, and things that are produced are products. List the reactants on the left side of a reaction arrow, separating each pair with a plus sign. Do the same for the products, but list them on the right side of the reaction arrow.

Q. Write a sentence that describes the following chemical reaction:

$$H_2O(l) + N_2O_3(g) \xrightarrow{\Delta} HNO_2(aq)$$

A. **Liquid water is heated with gaseous dinitrogen trioxide to produce an aqueous solution of nitrous acid.** First, figure out the names of the compounds. Next, note their states (liquid, solid, gas, or aqueous solution). Then observe what the compounds are actually doing — combining, decomposing, combusting, and so on. Finally, assemble all these observations into a sentence. Many sentence variations are correct, as long as they include these elements.

1. Write chemical equations for the following reactions:

a. Solid magnesium is heated with gaseous oxygen to form solid magnesium oxide.

b. Solid diboron trioxide reacts with solid magnesium to make solid boron and solid magnesium oxide.

Solve It

2. Write sentences describing the reactions summarized by the following chemical equations:

a. $S(s) + O_2(g) \rightarrow SO_2(g)$

b. $H_2(g) + O_2(g) \xrightarrow{Pt} H_2O(l)$

Solve It

Balancing Chemical Equations

The equations you read and write in the preceding section are *skeleton equations,* and they're perfectly adequate for a qualitative description of the reaction: What are the reactants, and what are the products? But if you look closely, you'll see that those equations just don't add up. As written, the mass of 1 mol of each of the reactants doesn't equal the mass of 1 mol of each of the products (see Chapter 7 for details on moles). The skeleton equations break the *law of conservation of mass,* which states that all the mass present at the beginning of a reaction must be present at the end. To be quantitatively accurate, these equations must be *balanced* so the masses of reactants and products are equal.

To balance an equation, you use *coefficients* to alter the number of moles of reactants and/or products so the mass on one side of the equation equals the mass on the other side. A *coefficient* is simply a number that precedes the symbol of an element or compound, multiplying the number of moles of that *entire* compound within the equation. Coefficients are different from *subscripts,* which multiply the number of atoms or groups within a compound. Consider the following:

$$4Cu(NO_3)_2$$

The number 4 that precedes the compound is a coefficient, indicating 4 mol of copper (II) nitrate. The subscripted 3 and 2 within the compound indicate that each nitrate contains three oxygen atoms and that there are two nitrate groups per copper ion. Coefficients and subscripts multiply to yield the total numbers of each atom present in the formula. In the preceding example, there are 4 atoms of copper present, 8 atoms of nitrogen, and 24 atoms of oxygen.

When you balance an equation, you change *only the coefficients.* Changing subscripts alters the chemical compounds themselves, and you can't do that. If your pencil were equipped with an electrical shocking device, that device would activate the moment you attempted to change a subscript while balancing an equation.

Here's a simple recipe for balancing equations:

1. **Given a skeleton equation (one that includes formulas for reactants and products), count up the number of each kind of atom on each side of the equation.**

   If you recognize any polyatomic ions and they're present in both the reactant and the product, you can count these as one whole group (as if they were their own form of element). See Chapter 6 for information on recognizing polyatomic ions.

2. **Use coefficients to balance the elements or polyatomic ions, one at a time.**

   For simplicity, start with those elements or ions that appear only once on each side.

3. **Check the equation to ensure that each element or ion is balanced.**

   Checking is important because you may have ping-ponged several times from reactants to products and back — there's plenty of opportunity for error.

4. **When you're sure the reaction is balanced, make sure it's in lowest terms.**

   For example, $4H_2(g) + 2O_2(g) \rightarrow 4H_2O(l)$ should be reduced to

   $$2H_2(g) + O_2(g) \rightarrow 2H_2O(l)$$

   Because you're reducing one of the coefficients to 1, you don't need to write it in the equation. If a compound doesn't have a coefficient written in front of it, you can assume a coefficient of 1.

Q. Balance the following equation:

$$Na(s) + Cl_2(g) \rightarrow NaCl(s)$$

A. **$2Na(s) + Cl_2(g) \rightarrow 2NaCl(s)$.** You can't change the subscripted 2 in the chlorine gas reactant, so you must add a coefficient of 2 to the sodium chloride product. This change requires you to balance the sodium reactant with another coefficient of 2.

3. Balance the following reactions:

a. $N_2(g) + H_2(g) \rightarrow NH_3(g)$

b. $C_3H_8(g) + O_2(g) \rightarrow CO_2(g) + H_2O(l)$

c. $Al(s) + O_2(g) \rightarrow Al_2O_3(s)$

d. $AgNO_3(aq) + Cu(s) \rightarrow Cu(NO_3)_2(aq) + Ag(s)$

Solve It

4. Balance the following reactions:

a. $Ag_2SO_4(aq) + AlCl_3(aq) \rightarrow AgCl(s)$
$$+ Al_2(SO_4)_3(aq)$$

b. $CH_4(g) + Cl_2(g) \rightarrow CH_2Cl_2(l) + HCl(g)$

c. $Cu(s) + HNO_3(aq) \rightarrow Cu(NO_3)_2(aq)$
$$+ NO(g) + H_2O(l)$$

d. $HCl(aq) + Ca(OH)_2(aq) \rightarrow CaCl_2(s) + H_2O(l)$

Solve It

5. Balance the following reactions:

a. $H_2C_2O_4(aq) + KOH(aq) \rightarrow K_2C_2O_4(aq) + H_2O(l)$

b. $P(s) + Br_2(l) \rightarrow PBr_3(g)$

c. $Pb(NO_3)_2(aq) + KI(aq) \rightarrow KNO_3(aq) + PbI_2(s)$

d. $Zn(s) + AgNO_3(aq) \rightarrow Zn(NO_3)_2(aq) + Ag(s)$

Solve It

Recognizing Reactions and Predicting Products

You can't begin to wrap your brain around the unimaginably large number of possible chemical reactions. That so many reactions can occur is a good thing, because they make things like life, the universe, and everything possible. From the perspective of a mere human brain trying to grok all these reactions, we have another bit of good news: A few categories of reactions pop up over and over again. After you see the very basic patterns in these categories, you'll be able to make sense of the majority of reactions out there.

The following sections describe five types of reactions that you'd do well to recognize (notice how their names tell you what happens in each reaction). By recognizing the patterns of these five types of reactions, you can often predict reaction products when given only a set of reactants. (*Note:* Figuring out the formulas of products often requires you to apply knowledge about how ionic and molecular compounds are put together. To review these concepts, see Chapters 5 and 6.)

Combination (synthesis)

In *combination* (sometimes called *synthesis*), two or more reactants combine to form a single product, following the general pattern

$$A + B \rightarrow AB$$

For example,

$$2Na(s) + Cl_2(g) \rightarrow 2NaCl(s)$$

The combining of elements to form compounds (like NaCl) is a particularly common kind of combination reaction. Here's another example:

$$2Ca(s) + O_2(g) \rightarrow 2CaO(s)$$

Compounds can also combine to form new compounds, such as in the combination of sodium oxide with water to form sodium hydroxide:

$$Na_2O(s) + H_2O(l) \rightarrow 2NaOH(aq)$$

Decomposition

In *decomposition,* a single reactant breaks down (decomposes) into two or more products, following the general pattern

$$AB \rightarrow A + B$$

For example,

$$2H_2O(l) \rightarrow 2H_2(g) + O_2(g)$$

Notice that combination and decomposition reactions are the same reaction in opposite directions.

Many decomposition reactions produce gaseous products, such as in the decomposition of carbonic acid into water and carbon dioxide:

$$H_2CO_3(aq) \rightarrow H_2O(l) + CO_2(g)$$

Single replacement (single displacement)

In a *single replacement* reaction (sometimes referred to as a *single displacement* reaction), a single, more reactive element or group replaces a less reactive element or group, following the general pattern

$$A + BC \rightarrow AC + B$$

For example,

$$Zn(s) + CuSO_4(aq) \rightarrow ZnSO_4(aq) + Cu(s)$$

Single replacement reactions in which metals replace other metals are especially common. Not all single replacement reactions occur as written, though. You sometimes need to refer to a chart called the *activity series* to determine whether such a reaction will take place. Table 8-2 presents the activity series. To determine whether a single replacement reaction will occur, compare the two metals in the reaction:

✔ If the metal that is single and not bonded is higher on the activity series than the metal that is bonded in the compound, the reaction will take place.

✔ If the metal by itself is not higher on the series, then no reaction will take place.

Table 8-2	Activity Series
Metal	*Notes*
Lithium	Most-reactive metals; react with cold water to form hydroxide and hydrogen gas.
Potassium	
Strontium	
Calcium	
Sodium	
Magnesium	React with hot water/acid to form oxides and hydrogen gas.
Aluminum	
Zinc	
Chromium	

(continued)

Table 8-2 (continued)

Metal	Notes
Iron	Replace hydrogen ion from dilute strong acids.
Cadmium	
Cobalt	
Nickel	
Tin	
Lead	
Hydrogen	Nonmetal, listed in reactive order.
Antimony	Combine directly with oxygen to form oxides.
Arsenic	
Bismuth	
Copper	
Mercury	Least-reactive metals; often found as free metals; oxides decompose easily.
Silver	
Palladium	
Platinum	
Gold	

Double replacement (double displacement)

Double replacement (also known as *double displacement*) is a special form of *metathesis reaction* — that is, a reaction in which two reacting species exchange bonds. Double replacement reactions tend to occur between ionic compounds in solution. In these reactions, cations (atoms or groups with a positive charge) from each reactant swap places to form ionic compounds with the opposing anions (atoms or groups with a negative charge), following the general pattern

$$AB + CD \rightarrow AD + CB$$

For example,

$$KCl(aq) + AgNO_3(aq) \rightarrow AgCl(s) + KNO_3(aq)$$

Of course, ions dissolved in solution move about freely, not as part of cation-anion complexes. So to allow double replacement reactions to progress, one of several things must occur:

✔ One of the product compounds must be insoluble so that it *precipitates* (forms an insoluble solid) out of solution after it forms.

✔ One of the products must be a gas that bubbles out of solution after it forms.

✔ One of the products must be a solvent molecule, such as H_2O, that separates from the ionic compounds after it forms.

Combustion

Oxygen is always a reactant in *combustion reactions,* which often release heat and light as they occur. Combustion reactions frequently involve hydrocarbon reactants (like propane, $C_3H_8(g)$, the gas used to fire up backyard grills) and yield carbon dioxide and water as products. For example,

$$C_3H_8(g) + 5\ O_2(g) \rightarrow 3CO_2(g) + 4H_2O(l)$$

Combustion reactions also include combination reactions between elements and oxygen, such as

$$S(s) + O_2(g) \rightarrow SO_2(g)$$

If the reactants include oxygen (O_2) and a hydrocarbon or an element, you're probably dealing with a combustion reaction. If the products are carbon dioxide and water, you're almost certainly dealing with a combustion reaction.

0. Predict and balance the following reaction:

$$Be(s) + O_2 \rightarrow$$

A. **2Be(s) + O$_2$ → 2BeO(s).** Although beryllium isn't on the metal activity series, you can make a pretty good prediction that this is a combination/combustion reaction. Why? First, you have two reactants. A single reactant would imply decomposition. The beryllium reactant is an element, not a compound, so you can rule out double replacement. The metal element reactant might make you consider single replacement, but there's no metal in oxygen, the other reactant, so there's no obvious replacement partner. So the most likely reaction is combination. When elements combine with oxygen, that's also a combustion reaction.

6. Complete and balance the following reactions:

    a. $C(s) + O_2(g) \rightarrow$

    b. $Mg(s) + I_2(g) \rightarrow$

    c. $Sr(s) + I_2(g) \rightarrow$

Solve It

7. Complete and balance the following reactions:

    a. $HI(g) \rightarrow$

    b. $H_2O_2(l) \rightarrow H_2O(l) +$

    c. $NaCl(s) \rightarrow$

Solve It

8. Complete and balance the following reactions:

    a. $Zn(s) + H_2SO_4(aq) \rightarrow$

    b. $Al(s) + HCl(aq) \rightarrow$

    c. $Li(s) + H_2O(l) \rightarrow$

Solve It

9. Complete and balance the following reactions:

    a. $Ca(OH)_2(aq) + HCl(aq) \rightarrow$

    b. $HNO_3(aq) + NaOH(aq) \rightarrow$

    c. $FeS(s) + H_2SO_4(aq) \rightarrow$

Solve It

Canceling Spectator Ions: Net Ionic Equations

Chemistry is often conducted in aqueous solutions. Soluble ionic compounds dissolve into their component ions, and these ions can react to form new products. In these kinds of reactions, sometimes only the cation or anion of a dissolved compound reacts. The other ion merely watches the whole affair, twiddling its charged thumbs in electrostatic boredom. These uninvolved ions are called *spectator ions*.

 Because spectator ions don't actually participate in the chemistry of a reaction, you don't need to include them in a chemical equation. Doing so leads to a needlessly complicated reaction equation, so chemists prefer to write *net ionic equations*, which omit the spectator ions. A net ionic equation doesn't include every component that may be present in a given beaker. Rather, it includes only those components that actually react.

 Here's a simple recipe for making net ionic equations of your own:

1. **Examine the starting equation to determine which ionic compounds are dissolved, as indicated by the (*aq*) symbol following the compound name.**

$$Zn(s) + HCl(aq) \rightarrow ZnCl_2(aq) + H_2(g)$$

2. **Rewrite the equation, explicitly separating dissolved ionic compounds into their component ions.**

$$Zn(s) + H^+(aq) + Cl^-(aq) \rightarrow Zn^{2+}(aq) + 2Cl^-(aq) + H_2(g)$$

Polyatomic ions don't break apart in solution, so be sure to familiarize yourself with the common ones (flip to Chapter 6 for details).

3. **Compare the reactant and product sides of the rewritten reaction and cross out the spectator ions.**

Any dissolved ions that appear in the same form on both sides are spectator ions. Cross out the spectator ions to produce a net reaction. If all reactants and products cross out, then no reaction will occur.

$$Zn(s) + H^+(aq) + \cancel{Cl^-(aq)} \rightarrow Zn^{2+}(aq) + \cancel{2Cl^-(aq)} + H_2(g)$$

The net reaction is

$$Zn(s) + H^+(aq) \rightarrow Zn^{2+}(aq) + H_2(g)$$

As written, the preceding reaction is imbalanced with respect to the number of hydrogen atoms and the amount of positive charge.

4. **Balance the net reaction for mass and charge.**

$$Zn(s) + 2H^+(aq) \rightarrow Zn^{2+}(aq) + H_2(g)$$

 If you want, you can balance the equation for mass and charge first (at Step 1). This way, when you cross out spectator ions at Step 3, you cross out equivalent numbers of ions. Either method produces the same net ionic equation in the end. Some people prefer to balance the starting reaction equation, but others prefer to balance the net reaction because it's a simpler equation.

Q. Generate a balanced net ionic equation for the following reaction:

$$CaCO_3(s) + 2HCl(aq) \rightarrow CaCl_2(aq) + H_2O(l) + CO_2(g)$$

A. $CaCO_3(s) + 2H^+(aq) \rightarrow Ca^{2+}(aq) + H_2O(l) + CO_2(g)$. Because HCl and $CaCl_2$ are listed as aqueous (aq), rewrite the equation, explicitly separating those compounds into their ionic components:

$$CaCO_3(s) + 2H^+(aq) + 2C^-(aq) \rightarrow Ca^{2+}(aq) + 2Cl^-(aq) + H_2O(l) + CO_2(g)$$

Next, cross out any components that appear in the same form on both sides of the equation. In this case, the chloride ions (Cl^-) are crossed out:

$$CaCO_3(s) + 2H^+(aq) + \underline{2Cl^-(aq)} \rightarrow Ca^{2+}(aq) + \underline{2Cl^-(aq)} + H_2O(l) + CO_2(g)$$

This leaves the net reaction:

$$CaCO_3(s) + 2H^+(aq) \rightarrow Ca^{2+}(aq) + H_2O(l) + CO_2(g)$$

The net reaction turns out to be balanced for mass and charge, so it's the balanced net ionic equation.

---

10. Generate balanced net ionic equations for the following reactions:

a. $LiOH(aq) + HI(aq) \rightarrow H_2O(l) + LiI(aq)$

b. $AgNO_3(aq) + NaCl(aq) \rightarrow AgCl(s) + NaNO_3(aq)$

c. $Pb(NO_3)_2(aq) + H_2SO_4(aq) \rightarrow PbSO_4(s) + HNO_3(aq)$

Solve It

---

11. Generate balanced net ionic equations for the following reactions:

a. $HCl(aq) + ZnS(aq) \rightarrow H_2S(g) + ZnCl_2(aq)$

b. $Ca(OH)_2(aq) + H_3PO_4(aq) \rightarrow Ca_3(PO_4)_2(aq) + H_2O(l)$

c. $(NH_4)_2 S(aq) + Co(NO_3)_2(aq) \rightarrow CoS(s) + NH_4NO_3(aq)$

Solve It

Answers to Questions on Chemical Equations

Chemistry is about action, the breaking and making of bonds. You've read the chapter, and you've answered the questions. Now check your answers to see whether the chemistry concepts in this chapter acted on your brain.

1 Don't forget the symbols indicating state and whether any heat or catalysts were added.

a. $Mg(s) + O_2(g) \xrightarrow{\Delta} MgO(s)$

b. $B_2O_3(s) + Mg(s) \rightarrow B(s) + MgO(s)$

2 Here are the sentences describing the provided reactions:

a. Solid sulfur and gaseous oxygen react to produce the gas sulfur dioxide.

b. Hydrogen gas reacts with oxygen gas in the presence of platinum to produce liquid water.

3 Here are the balanced reactions:

a. $N_2(g) + 3H_2(g) \rightarrow 2NH_3(g)$

b. $C_3H_8(g) + 5\ O_2(g) \rightarrow 3CO_2(g) + 4H_2O(l)$

c. $4Al(s) + 3\ O_2(g) \rightarrow 2Al_2O_3(s)$

d. $2AgNO_3(aq) + Cu(s) \rightarrow Cu(NO_3)_2(aq) + 2Ag(s)$

4 More balanced reactions:

a. $3Ag_2SO_4(aq) + 2AlCl_3(aq) \rightarrow 6AgCl(s) + Al_2(SO_4)_3(aq)$

b. $CH_4(g) + 2Cl_2(g) \rightarrow CH_2Cl_2(l) + 2HCl(g)$

c. $3Cu(s) + 8HNO_3(aq) \rightarrow 3Cu(NO_3)_2(aq) + 2NO(g) + 4H_2O(l)$

d. $2HCl(aq) + Ca(OH)_2(aq) \rightarrow CaCl_2(s) + 2H_2O(l)$

5 Still more balanced reactions:

a. $H_2C_2O_4(aq) + 2KOH(aq) \rightarrow K_2C_2O_4(aq) + 2H_2O(l)$

b. $2P(s) + 3Br_2(l) \rightarrow 2PBr_3(g)$

c. $Pb(NO_3)_2(aq) + 2KI(aq) \rightarrow 2KNO_3(aq) + PbI_2(s)$

d. $Zn(s) + 2AgNO_3(aq) \rightarrow Zn(NO_3)_2(aq) + 2Ag(s)$

6 After completing the reaction, forgetting to balance it, too, is easy. Of course, proper balancing means that you have to pay attention to the amount of each atom in the product compounds. These reactions are all combination reactions:

a. $C(s) + O_2(g) \rightarrow CO_2(g)$

b. $Mg(s) + I_2(g) \rightarrow MgI(s)$

c. $Sr(s) + I_2(g) \rightarrow SrI_2(s)$

7 Here are some more completed, balanced reactions. All of these are decomposition reactions:

a. $2HI(g) \rightarrow H_2 + I_2$

b. $2H_2O_2(l) \rightarrow 2H_2O(l) + O_2(g)$

c. $2NaCl(s) \rightarrow 2Na(s) + Cl_2(g)$

8   Here are the completed, balanced versions of a series of single replacement reactions:

     a. $Zn(s) + H_2SO_4(aq) \rightarrow ZnSO_4(aq) + H_2(g)$

     b. $2Al(s) + 6HCl(aq) \rightarrow 2AlCl_3(aq) + 3H_2(g)$

     c. $2Li(s) + 2H_2O(l) \rightarrow 2LiOH(aq) + H_2(g)$

9   This final set, completed and balanced, is composed of double replacement reactions:

     a. $Ca(OH)_2(aq) + 2HCl(aq) \rightarrow CaCl_2(aq) + 2H_2O(l)$

     b. $HNO_3(aq) + NaOH(aq) \rightarrow H_2O(l) + NaNO_3(aq)$

     c. $FeS(s) + H_2SO_4(aq) \rightarrow H_2S(g) + FeSO_4(aq)$

10   The following answers show the original reaction, the expanded reaction, the expanded reaction with spectator ions crossed out, and the final balanced reaction.

     a. $$LiOH(aq) + HI(aq) \rightarrow H_2O(l) + LiI(aq)$$

$$Li^+(aq) + OH^-(aq) + H^+(aq) + I^-(aq) \rightarrow H_2O(l) + Li^+(aq) + I^-(aq)$$

$$\cancel{Li^+(aq)} + OH^-(aq) + H^+(aq) + \cancel{I^-(aq)} \rightarrow H_2O(l) + \cancel{Li^+(aq)} + \cancel{I^-(aq)}$$

$$OH^-(aq) + H^+(aq) \rightarrow H_2O(l)$$

     b. $$AgNO_3(aq) + NaCl(aq) \rightarrow AgCl(s) + NaNO_3(aq)$$

$$Ag^+(aq) + NO_3^-(aq) + Na^+(aq) + Cl^-(aq) \rightarrow AgCl(s) + Na^+(aq) + NO_3^-(aq)$$

$$Ag^+(aq) + \cancel{NO_3^-(aq)} + \cancel{Na^+(aq)} + Cl^-(aq) \rightarrow AgCl(s) + \cancel{Na^+(aq)} + \cancel{NO_3^-(aq)}$$

$$Ag^+(aq) + Cl^-(aq) \rightarrow AgCl(s)$$

     c. $$Pb(NO_3)_2(aq) + H_2SO_4(aq) \rightarrow PbSO_4(s) + HNO_3(aq)$$

$$Pb^{2+}(aq) + 2NO_3^-(aq) + 2H^+(aq) + SO_4^{2-}(aq) \rightarrow PbSO_4(s) + H^+(aq) + NO_3^-(aq)$$

$$Pb^{2+}(aq) + \cancel{2NO_3^-(aq)} + \cancel{2H^+(aq)} + SO_4^{2-}(aq) \rightarrow PbSO_4(s) + \cancel{H^+(aq)} + \cancel{NO_3^-(aq)}$$

$$Pb^{2+}(aq) + SO_4^{2-}(aq) \rightarrow PbSO_4(s)$$

11   Here are the answers for the second batch of net ionic equation questions, again showing the original reaction, the expanded reaction, the expanded reaction with spectator ions crossed out, and the final balanced reaction.

     a. $$HCl(aq) + ZnS(aq) \rightarrow H_2S(g) + ZnCl_2(aq)$$

$$H^+(aq) + Cl^-(aq) + Zn^{2+}(aq) + S^{2-}(aq) \rightarrow H_2S(g) + Zn^{2+}(aq) + 2Cl^-(aq)$$

$$H^+(aq) + \cancel{Cl^-(aq)} + \cancel{Zn^{2+}(aq)} + S^{2-}(aq) \rightarrow H_2S(g) + \cancel{Zn^{2+}(aq)} + \cancel{2Cl^-(aq)}$$

$$2H^+(aq) + S^{2-}(aq) \rightarrow H_2S(g)$$

     b. $$Ca(OH)_2(aq) + H_3PO_4(aq) \rightarrow Ca_3(PO_4)_2(aq) + H_2O(l)$$

$$Ca^{2+}(aq) + 2OH^-(aq) + 3H^+(aq) + PO_4^{3-}(aq) \rightarrow 3Ca^{2+}(aq) + 2PO_4^{3-}(aq) + H_2O(l)$$

$$\cancel{Ca^{2+}(aq)} + 2OH^-(aq) + 3H^+(aq) + \cancel{PO_4^{3-}(aq)} \rightarrow \cancel{3Ca^{2+}(aq)} + \cancel{2PO_4^{3-}(aq)} + H_2O(l)$$

$$OH^-(aq) + H^+(aq) \rightarrow H_2O(l)$$

     c. $$(NH_4)_2S(aq) + Co(NO_3)_2(aq) \rightarrow CoS(s) + NH_4NO_3(aq)$$

$$2NH_4^+(aq) + S^{2-}(aq) + Co^{2+}(aq) + 2NO_3^-(aq) \rightarrow CoS(s) + NH_4^+(aq) + NO_3^-(aq)$$

$$\cancel{2NH_4^+(aq)} + S^{2-}(aq) + Co^{2+}(aq) + \cancel{2NO_3^-(aq)} \rightarrow CoS(s) + \cancel{NH_4^+(aq)} + \cancel{NO_3^-(aq)}$$

$$S^{2-}(aq) + Co^{2+}(aq) \rightarrow CoS(s)$$

Chapter 9

Putting Stoichiometry to Work

. . . .

In This Chapter

▶ Doing conversions: Mole-mole, mole-particles, mole-volume, and mole-mass

▶ Figuring out what happens when one reagent runs out before the others

▶ Using percent yield to determine the efficiency of reactions

. . . .

Stoichiometry. Such a complicated word for such a simple idea. The Greek roots of the word mean "measuring elements," which doesn't sound nearly as intimidating. Moreover, the ancient Greeks couldn't tell an ionic bond from an Ionic column, so just how technical and scary could stoichiometry really be? Simply stated, *stoichiometry* is the quantitative relationship between components of chemical substances. In compound formulas and reaction equations, you express stoichiometry by using subscripted numbers and coefficients.

If you arrived at this chapter by first wandering through Chapters 6, 7, and 8, then you've already had breakfast, lunch, and an afternoon snack with stoichiometry. If you bypassed the aforementioned chapters, then you haven't eaten all day. Either way, it's time for dinner. Please pass the coefficients.

We've rounded molar masses to the hundredths place before doing the calculations. Answers have been rounded according to the rules for significant figures (see Chapter 1 for details).

Using Mole-Mole Conversions from Balanced Equations

Mass and energy are conserved. It's the law. Unfortunately, this means that there's no such thing as a free lunch or any other type of free meal. Ever. On the other hand, the conservation of mass makes it possible to predict how chemical reactions will turn out.

Chapter 8 describes why chemical reaction equations should be balanced for equal mass in reactants and products. You balance an equation by adjusting the coefficients that precede reactant and product compounds within the equation. Balancing equations can seem like a chore, like taking out the trash. But a balanced equation is far better than any collection of coffee grounds and orange peels, because such an equation is a useful tool. After you've got a balanced equation, you can use the coefficients to build *mole-mole conversion factors*. These kinds of conversion factors tell you how much of any given product you get by reacting any given amount of reactant. This is one of those calculations that makes chemists particularly useful, so they needn't get by on looks and charm alone. (For more about moles, see Chapter 7.)

Consider the following balanced equation for generating ammonia from nitrogen and hydrogen gases:

$$N_2(g) + 3H_2(g) \rightarrow 2NH_3(g)$$

Industrial chemists around the globe perform this reaction, humorlessly fixating on how much ammonia product they'll end up with at the end of the day. (In fact, clever methods for improving the rate and yield of this reaction garnered Nobel Prizes for two German gentlemen, die Herren Haber und Bosch.) In any event, how are chemists to judge how closely their reactions have approached completion? The heart of the answer lies in a balanced equation and the mole-mole conversion factors that spring from it.

For every mole of nitrogen reactant, a chemist expects 2 moles of ammonia product. Similarly, for every 3 moles of hydrogen reactant, the chemist expects 2 moles of ammonia product. These expectations are based on the coefficients of the balanced equation and are expressed as mole-mole conversion factors as shown in Figure 9-1.

Figure 9-1: Building mole-mole conversion factors from a balanced equation.

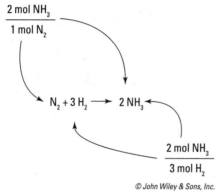

© John Wiley & Sons, Inc.

Q. How many moles of ammonia can be expected from the reaction of 278 mol of N_2 gas?

A. **556 mol of ammonia.** Begin with your known quantity, the 278 mol of nitrogen that's to be reacted. Multiply that quantity by the mole-mole conversion factor that relates moles of nitrogen to moles of ammonia. Write the conversion factor so that *mol NH$_3$* is on top and *mol N$_2$* is on the bottom. That way, the *mol N$_2$* units cancel, leaving you with the desired units, *mol NH$_3$*. The numbers you put in front of the units for the conversion factor come directly from the coefficients in the balanced chemical equation.

$$(278 \text{ mol } N_2) \left(\frac{2 \text{ mol } NH_3}{1 \text{ mol } N_2} \right) = 556 \text{ mol } NH_3$$

1. One source of hydrogen gas is the electrolysis of water, in which electricity is passed through water to break hydrogen-oxygen bonds, yielding hydrogen and oxygen gases:

$$2H_2O(l) \rightarrow 2H_2(g) + O_2(g)$$

a. How many moles of hydrogen gas result from the electrolysis of 78.4 mol of water?

b. How many moles of water are required to produce 905 mol of hydrogen?

c. Running the electrolysis reaction in reverse constitutes the combustion of hydrogen. How many moles of oxygen are required to combust 84.6 mol of hydrogen?

Solve It

2. Aluminum reacts with copper (II) sulfate to produce aluminum sulfate and copper, as summarized by this skeleton equation:

$$Al(s) + CuSO_4(aq) \rightarrow Al_2(SO_4)_3(aq) + Cu(s)$$

a. Balance the equation.

b. How many moles of aluminum are needed to react with 10.38 mol of copper (II) sulfate?

c. How many moles of copper are produced if 2.08 mol of copper (II) sulfate react with aluminum?

d. How many moles of copper (II) sulfate are needed to produce 0.96 mol of aluminum sulfate?

e. How many moles of aluminum are needed to produce 20.01 mol of copper?

Solve It

3. Solid iron reacts with solid sulfur to form iron (III) sulfide:

$$Fe(s) + S(s) \rightarrow Fe_2S_3(s)$$

a. Balance the equation.

b. How many moles of sulfur are needed to react with 6.2 mol of iron?

c. How many moles of iron (III) sulfide are produced from 10.6 mol of iron?

d. How many moles of iron (III) sulfide are produced from 3.5 mol of sulfur?

Solve It

4. Ethane combusts to form carbon dioxide and water:

$$C_2H_6(g) + O_2 \rightarrow CO_2(g) + H_2O(g)$$

a. Balance the equation.

b. 15.4 mol of ethane produces how many moles of carbon dioxide?

c. How many moles of ethane does it take to produce 293 mol of water?

d. How many moles of oxygen are required to combust 0.178 mol of ethane?

Solve It

Putting Moles at the Center: Conversions Involving Particles, Volumes, and Masses

The mole is the beating heart of stoichiometry, the central unit through which other quantities flow. Real-life chemists don't have magic mole vision, however. A chemist can't look at a pile of potassium chloride crystals, squint her eyes, and proclaim, "That's 0.539 moles of salt." Real *reagents* (reactants) tend to be measured in units of mass or volume. Real products are measured in the same way. So you need to be able to use mole-mass, mole-volume, and mole-particle conversion factors to translate between these different dialects of counting. Figure 9-2 summarizes the interrelationship among all these things and serves as a flow-chart for problem-solving. All roads lead to and from the mole.

Figure 9-2:
A problem-solving flowchart showing the use of mole-mole, mole-mass, mole-volume, and mole-particle conversion factors.

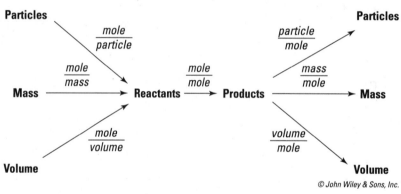

© John Wiley & Sons, Inc.

If you look at Figure 9-2, you can see that it isn't possible to convert directly between the mass of one substance and the mass of another substance. You must convert to moles and then use the mole-mole conversion factor before converting to the mass of a new substance. The same can be said for conversions from the particles or volume of one substance to that of another substance. The mole is always the intermediary you use for the conversion.

Do all your calculations in one long stoichiometric problem, saving any rounding for the end. When you have a string of conversion factors, you're far better off calculating everything in one step instead of breaking up a problem into multiple steps where you type a calculation into your calculator, hit *enter,* and then move on to the next step. Doing that leads to a greater degree of error.

Q. Calcium carbonate decomposes to produce solid calcium oxide and carbon dioxide gas according to the following reaction. Answer each part of the question, assuming that 10.0 g of calcium carbonate decomposes.

$$CaCO_3(s) \rightarrow CaO(s) + CO_2(g)$$

a. How many grams of calcium oxide are produced?

b. At standard temperature and pressure (STP), how many liters of carbon dioxide are produced?

A. Here are the answers:

a. 5.61 g CaO. First, convert 10.0 g of calcium carbonate to moles of calcium carbonate by using the molar mass of calcium carbonate (100.09 g/mol) as a conversion factor (see Chapter 7 for more about molar mass). To determine the grams of calcium oxide produced, you must then convert from moles of calcium carbonate to moles of calcium oxide. Keep in mind that you take the number of moles for the mole-mole conversion from the coefficients in the balanced chemical equation (see Chapter 8). Then convert from moles of calcium oxide to grams of calcium oxide by using the molar mass of calcium oxide (56.08 g/mol) as a conversion factor:

$$(10.0 \text{ g CaCO}_3)\left(\frac{1 \text{ mol CaCO}_3}{100.09 \text{ g CaCO}_3}\right)\left(\frac{1 \text{ mol CaO}}{1 \text{ mol CaCO}_3}\right)\left(\frac{56.08 \text{ g CaO}}{1 \text{ mol CaO}}\right) = 5.61 \text{ g CaO}$$

b. 2.24 L CO$_2$. To determine the liters of carbon dioxide produced, follow the initial mass-mole conversion with a mole-mole conversion to find the moles of carbon dioxide produced. Then convert from moles of carbon dioxide to liters by using the fact that at STP, each mole of gas occupies 22.4 L:

$$(10.0 \text{ g CaCO}_3)\left(\frac{1 \text{ mol CaCO}_3}{100.09 \text{ g CaCO}_3}\right)\left(\frac{1 \text{ mol CO}_2}{1 \text{ mol CaCO}_3}\right)\left(\frac{22.4 \text{ L CO}_2}{1 \text{ mol CO}_2}\right) = 2.24 \text{ L CO}_2$$

Notice how both calculations require you to first convert to moles and then perform a mole-mole conversion using stoichiometry from the reaction equation. Then you convert to the desired units. Both solutions consist of a chain of conversion factors, each factor bringing the units one step closer to those needed in the answer.

5. Hydrogen peroxide decomposes into oxygen gas and liquid water:

$$H_2O_2(l) \rightarrow H_2O(l) + O_2(g)$$

a. Balance the equation.

b. How many grams of water are produced when 2.94×10^{24} molecules of hydrogen peroxide decompose? (Remember that molecules are particles.)

c. What is the volume of oxygen produced at STP when 32.9 g of hydrogen peroxide decomposes?

Solve It

6. Dinitrogen trioxide gas reacts with liquid water to produce nitrous acid:

$$N_2O_3(g) + H_2O(l) \rightarrow HNO_2(aq)$$

a. Balance the equation.

b. At STP, how many liters of dinitrogen trioxide react to produce 36.98 g of dissolved nitrous acid?

c. How many molecules of water react with 17.3 L of dinitrogen trioxide at STP?

Solve It

7. Lead (II) chloride reacts with chlorine to produce lead (IV) chloride:

$$PbCl_2(s) + Cl_2(g) \rightarrow PbCl_4(l)$$

a. What volume of chlorine reacts to convert 50.0 g of lead (II) chloride entirely into product?

b. How many formula units (particles of an ionic compound) of lead (IV) chloride result from the reaction of 13.71 g of lead (II) chloride?

c. How many grams of lead (II) chloride react to produce 84.8 g of lead (IV) chloride?

Solve It

Limiting Your Reagents

In real-life chemical reactions, not all of the reactants present convert into product. That would be perfect and convenient. Does that sound like real life to you? More typically, one reagent is completely used up, and others are left in excess, perhaps to react another day.

The situation resembles that of a horde of Hollywood hopefuls lined up for a limited number of slots as extras in a film. Only so many eager faces react with an available slot to produce a happily (albeit pitifully) employed actor. The remaining actors are in excess, muttering quietly all the way back to their jobs as waiters. In this scenario, the slots are the limiting reagent.

Those standing in line demand to know, "How many slots are there?" With this key piece of data, they can deduce how many in their huddled mass will end up with a gig. Or they can figure out how many will continue to waste their film school degrees serving penne with basil and goat cheese to chemists on vacation.

Chemists demand to know, "Which reactant will run out first?" In other words, which reactant is the limiting reagent? Knowing that information allows them to deduce how much product they can expect, based on how much of the limiting reagent they've put into the reaction. Also, identifying the limiting reagent allows them to calculate how much of the excess reagent they'll have left over when all the smoke clears. Either way, the first step is to figure out which is the limiting reagent.

In any chemical reaction, you can simply pick one reagent as a candidate for the limiting reagent, calculate how many moles of that reagent you have, and then calculate how many grams of the other reagent you'd need to react both to completion. You'll discover one of two things. Either you have an excess of the first reagent, or you have an excess of the second reagent. The one you have in excess is the *excess reagent.* The one that isn't in excess is the *limiting reagent.* When you know which reactant is limiting, you can use that info to solve any stoichiometry problems involving that reaction. Why? You must use the limiting reactant for calculations because it runs out first and governs how much of the products you'll create in the reaction. If you used the excess for a calculation, you'd end up with a much larger amount of product than you should.

Q. Ammonia reacts with oxygen to produce nitrogen monoxide and liquid water:

$$NH_3(g) + O_2(g) \rightarrow NO(g) + H_2O(l)$$

a. Balance the equation (as we explain in Chapter 8).

b. Determine the limiting reagent if 100 g of ammonia and 100 g of oxygen are present at the beginning of the reaction.

c. What is the excess reagent, and how many grams of the excess reagent will remain when the reaction reaches completion?

d. How many grams of nitrogen monoxide will be produced if the reaction goes to completion?

e. How many grams of water will be produced if the reaction goes to completion?

A. Here's the solution:

a. Before doing anything else, you must have a balanced reaction equation. Don't waste good thought on an unbalanced equation. The balanced form of the given equation is

$$4NH_3(g) + 5O_2(g) \rightarrow 4NO(g) + 6H_2O(l)$$

b. Oxygen is the limiting reagent. Two candidates, NH_3 and O_2, vie for the status of limiting reagent. You start with 100 g of each, which corresponds to some number of moles of each. Furthermore, you can tell from the coefficients in the balanced equation this reaction requires 4 mol of ammonia for every 5 mol of oxygen gas.

To find the limiting reactant, you simply need to perform a mass-to-mass (gram-to-gram) calculation from one reactant to the other (see Chapter 7 for more about molar mass). This allows you to see which reactant runs out first. You can start with either reactant and convert to mass of the other. In this example, we start with ammonia:

$$(100 \text{ g NH}_3) \left(\frac{1 \text{ mol NH}_3}{17.04 \text{ g NH}_3} \right) \left(\frac{5 \text{ mol O}_2}{4 \text{ mol NH}_3} \right) \left(\frac{32.00 \text{ g O}_2}{1 \text{ mol O}_2} \right) = 235 \text{ g O}_2$$

The calculation reveals that you'd need 235 g of oxygen gas to completely react with 100 g of ammonia. But you have only 100 g of oxygen. You'll run out of oxygen before you run out of ammonia, so oxygen is the limiting reagent.

c. The excess reagent is ammonia, and 57.5 g of ammonia will remain when the reaction reaches completion. To calculate how many grams of ammonia will be left at the end of the reaction, assume that all 100 g of oxygen react:

$$(100 \text{ g O}_2) \left(\frac{1 \text{ mol O}_2}{32.00 \text{ g O}_2} \right) \left(\frac{4 \text{ mol NH}_3}{5 \text{ mol O}_2} \right) \left(\frac{17.04 \text{ g NH}_3}{1 \text{ mol NH}_3} \right) = 42.5 \text{ g NH}_3$$

This calculation shows that 42.5 g of the original 100 g of ammonia will react before the limiting reagent is expended. So 57.5 g of ammonia will remain in excess (just subtract 42.5 from 100).

d. 75 g of nitrogen monoxide will be produced. This problem asks how much of a product is produced. For this calculation, you must begin with the limiting reactant. To determine the grams of nitrogen monoxide that are generated by the complete reaction of oxygen, start with the assumption that all 100 g of the oxygen react:

$$(100 \text{ g O}_2) \left(\frac{1 \text{ mol O}_2}{32.00 \text{ g O}_2} \right) \left(\frac{4 \text{ mol NO}}{5 \text{ mol O}_2} \right) \left(\frac{30.01 \text{ g NO}}{1 \text{ mol NO}} \right) = 75.0 \text{ g NO}$$

e. 67.5g of water will be produced. Again, assume that all 100 g of the oxygen react in order to determine how many grams of water are produced:

$$(100 \text{ g O}_2) \left(\frac{1 \text{ mol O}_2}{32.00 \text{ g O}_2} \right) \left(\frac{6 \text{ mol H}_2O}{5 \text{ mol O}_2} \right) \left(\frac{18.02 \text{ g H}_2O}{1 \text{ mol H}_2O} \right) = 67.5 \text{ g H}_2O$$

8. Iron (III) oxide reacts with carbon monoxide to produce iron and carbon dioxide:

$$Fe_2O_3(s) + 3CO(g) \rightarrow 2Fe(s) + 3CO_2(g)$$

a. What is the limiting reagent if 50 g of iron (III) oxide and 67 g of carbon monoxide are present at the beginning of the reaction?

b. What is the excess reagent, and how many grams of it will remain after the reaction proceeds to completion?

c. How many grams of each product should be expected if the reaction goes to completion?

Solve It

9. Solid sodium reacts (violently) with water to produce sodium hydroxide and hydrogen gas:

$$2Na(s) + 2H_2O(l) \rightarrow 2NaOH(aq) + H_2(g)$$

a. What is the limiting reagent if 25 g of sodium and 40.2 g of water are present at the beginning of the reaction?

b. What is the excess reagent, and how many grams of it will remain after the reaction has gone to completion?

c. How many grams of sodium hydroxide and how many liters of hydrogen gas (at STP) should be expected if the reaction goes to completion?

Solve It

10. Aluminum reacts with chlorine gas to produce aluminum chloride:

$$2Al(s) + 3Cl_2(g) \rightarrow 2AlCl_3(s)$$

a. What is the limiting reagent if 29.3 g of aluminum and 34.6 L of chlorine gas (at STP) are present at the beginning of the reaction?

b. What is the excess reagent, and how many grams (or liters at STP) of it will remain after the reaction has gone to completion?

c. How many grams of aluminum chloride should be expected if the reaction goes to completion?

Solve It

Counting Your Chickens after They've Hatched: Percent Yield Calculations

In a way, reactants have it easy. Maybe they'll make something of themselves and actually react. Or maybe they'll just lean against the inside of the beaker, flipping through a back issue of *People* magazine and sipping a caramel macchiato.

Chemists don't have it so easy. Someone is paying them to do reactions. That someone doesn't have time or money for excuses about loitering reactants. So you, as a fresh-faced chemist, have to be concerned with just how completely your reactants react to form products. To compare the amount of product obtained from a reaction with the amount that should have been obtained, chemists use *percent yield*. You determine percent yield with the following formula:

$$\text{Percent Yield} = \left(\frac{\text{Actual Yield}}{\text{Theoretical Yield}} \right) \times 100$$

Lovely, but what is an actual yield, and what is a theoretical yield? An *actual yield* is, well, the amount of product actually produced by the reaction in a lab or as told to you in the chemistry problem. A *theoretical yield* is the amount of product that could've been produced had everything gone perfectly, as described by theory if every single atom of reactants worked together perfectly. The theoretical yield is what you calculate when you do a calculation on paper (like in this chapter) or before you do a reaction in a lab.

The actual yield will always be less than the theoretical yield because no chemical reaction ever reaches 100 percent completion. In a lab setting, there's always some amount of error, whether it's big or small.

Q. Calculate the percent yield of sodium sulfate in the following scenario: 32.18 g of sulfuric acid reacts with excess sodium hydroxide to produce 37.91 g of sodium sulfate.

$$H_2SO_4(aq) + 2NaOH(aq) \rightarrow 2H_2O(l) + Na_2SO_4(aq)$$

A. **81.37% is the percent yield.** The question clearly notes that sodium hydroxide is the excess reagent. (*Tip:* You always can ignore a reactant if the problem says it's in excess. That's like a big this-one-isn't-important sign in the problem.) So sulfuric acid is the limiting reagent and is the reagent you should use to calculate the theoretical yield:

$$\left(32.18 \text{ g } H_2SO_4 \right) \left(\frac{1 \text{ mol } H_2SO_4}{98.08 \text{ g } H_2SO_4} \right) \left(\frac{1 \text{ mol } Na_2SO_4}{1 \text{ mol } H_2SO_4} \right) \left(\frac{142.04 \text{ g } Na_2SO_4}{1 \text{ mol } Na_2SO_4} \right) = 46.59 \text{ g } Na_2SO_4$$

Theory predicts that 46.59 g of sodium sulfate product is possible if the reaction proceeds perfectly and to completion. But the question states that the actual yield is only 37.91 g of sodium sulfate. With these two pieces of information, you can calculate the percent yield using the formula given at the start of this section:

$$\text{Percent Yield} = \left(\frac{37.91 \text{ g}}{46.59 \text{ g}} \right) \times 100 = 81.37\%$$

---

11. Sulfur dioxide reacts with liquid water to produce sulfurous acid:

$$SO_2(g) + H_2O(l) \rightarrow H_2SO_3(aq)$$

a. What is the percent yield if 19.07 g of sulfur dioxide reacts with excess water to produce 21.61 g of sulfurous acid?

b. When 8.11 g of water reacts with excess sulfur dioxide, 27.59 g of sulfurous acid is produced. What is the percent yield?

Solve It

12. Liquid hydrazine is a component of some rocket fuels. Hydrazine combusts to produce nitrogen gas and water:

$$N_2H_4(l) + O_2(g) \rightarrow N_2(g) + 2H_2O(g)$$

a. If the percent yield of a combustion reaction in the presence of 23.4 g of N_2H_4 (and excess oxygen) is 98%, how many liters of nitrogen (at STP) are produced?

b. What is the percent yield if 84.8 g of N_2H_4 reacts with 54.7 g of oxygen gas to produce 51.33 g of water?

Solve It

Answers to Questions on Stoichiometry

Balanced equations, diverse conversion factors, limiting reagents, and percent yield calculations all now tremble before you. Probably. Be sure your powers over them are as breathtaking as they should be by checking your answers.

1 In the following equations, keep in mind that the number you use for moles is the coefficient for the compound taken from the balanced chemical reaction:

a. 78.4 mol H$_2$.

$$(78.4 \text{ mol H}_2\text{O}) \left(\frac{2 \text{ mol H}_2}{2 \text{ mol H}_2\text{O}} \right) = 78.4 \text{ mol H}_2$$

b. 905 mol H$_2$O.

$$(905 \text{ mol H}_2) \left(\frac{2 \text{ mol H}_2\text{O}}{2 \text{ mol H}_2} \right) = 905 \text{ mol H}_2\text{O}$$

c. 42.3 mol O$_2$.

$$(84.6 \text{ mol H}_2) \left(\frac{1 \text{ mol O}_2}{2 \text{ mol H}_2} \right) = 42.3 \text{ mol O}_2$$

2 Before attempting to do any calculations, you must balance the equation as shown in part (a) so that you can use the coefficients to do your conversions (see Chapter 8 for info on balancing equations).

a. The balanced equation is

$$2Al(s) + 3CuSO_4(aq) \rightarrow Al_2(SO_4)_3(aq) + 3Cu(s)$$

b. 6.920 mol Al.

$$(10.38 \text{ mol CuSO}_4) \left(\frac{2 \text{ mol Al}}{3 \text{ mol CuSO}_4} \right) = 6.920 \text{ mol Al}$$

c. 2.08 mol Cu.

$$(2.08 \text{ mol CuSO}_4) \left(\frac{3 \text{ mol Cu}}{3 \text{ mol CuSO}_4} \right) = 2.08 \text{ mol Cu}$$

d. 2.9 mol CuSO$_4$.

$$(0.96 \text{ mol Al}_2(\text{SO}_4)_3) \left(\frac{3 \text{ mol CuSO}_4}{1 \text{ mol Al}_2(\text{SO}_4)_3} \right) = 2.9 \text{ mol CuSo}_4$$

e. 13.34 mol Al.

$$(20.01 \text{ mol Cu}) \left(\frac{2 \text{ mol Al}}{3 \text{ mol Cu}} \right) = 13.34 \text{ mol Al}$$

3 Balance the equation first. Then go on to the calculations for the remaining questions.

a. The balanced equation is

$$2Fe(s) + 3S(s) \rightarrow Fe_2S_3(s)$$

b. 9.3 mol S.

$$(6.2 \text{ mol Fe}) \left(\frac{3 \text{ mol S}}{2 \text{ mol Fe}} \right) = 9.3 \text{ mol S}$$

c. 5.30 mol Fe$_2$S$_3$.

$$(10.6 \text{ mol Fe}) \left(\frac{1 \text{ mol Fe}_2\text{S}_3}{2 \text{ mol Fe}} \right) = 5.30 \text{ mol Fe}_2\text{S}_3$$

d. 1.2 mol Fe$_2$S$_3$.

$$(3.5 \text{ mol S}) \left(\frac{1 \text{ mol Fe}_2\text{S}_3}{3 \text{ mol S}} \right) = 1.2 \text{ mol Fe}_2\text{S}_3$$

4 As always, begin by balancing the equation.

a. The balanced equation is

$$2C_2H_6(g) + 7O_2 \rightarrow 4CO_2(g) + 6H_2O(g)$$

b. 30.8 mol CO$_2$.

$$(15.4 \text{ mol C}_2\text{H}_6) \left(\frac{4 \text{ mol CO}_2}{2 \text{ mol C}_2\text{H}_6} \right) = 30.8 \text{ mol CO}_2$$

c. 97.7 mol C$_2$H$_6$.

$$(293 \text{ mol H}_2\text{O}) \left(\frac{2 \text{ mol C}_2\text{H}_6}{6 \text{ mol H}_2\text{O}} \right) = 97.7 \text{ mol C}_2\text{H}_6$$

d. 0.623 mol O$_2$.

$$(0.178 \text{ mol C}_2\text{H}_6) \left(\frac{7 \text{ mol O}_2}{2 \text{ mol C}_2\text{H}_6} \right) = 0.623 \text{ mol O}_2$$

5 Without a balanced equation (see Chapter 8), nothing else is possible.

a. The balanced equation is

$$2H_2O_2(l) \rightarrow 2H_2O(l) + O_2(g)$$

b. 87.9 g H$_2$O.

$$(2.94 \times 10^{24} \text{ molec H}_2\text{O}_2) \left(\frac{1 \text{ mol H}_2\text{O}_2}{6.022 \times 10^{23} \text{ molec H}_2\text{O}_2} \right)$$

$$\left(\frac{2 \text{ mol H}_2\text{O}}{2 \text{ mol H}_2\text{O}_2} \right) \left(\frac{18.02 \text{ g H}_2\text{O}}{1 \text{ mol H}_2\text{O}} \right) = 87.9 \text{ g H}_2\text{O}$$

c. 10.8 L O_2.

$$(32.9 \text{ g } H_2O_2)\left(\frac{1 \text{ mol } H_2O_2}{34.02 \text{ g } H_2O_2}\right)\left(\frac{1 \text{ mol } O_2}{2 \text{ mol } H_2O_2}\right)\left(\frac{22.4 \text{ L } O_2}{1 \text{ mol } O_2}\right) = 10.8 \text{ L } O_2$$

6 The conversion factors you need to do the calculations require a balanced equation, so do the balancing first.

a. The balanced equation is

$$N_2O_3(g) + H_2O(l) \rightarrow 2HNO_2(aq)$$

b. 8.81 L N_2O_3.

$$(36.98 \text{ g } HNO_2)\left(\frac{1 \text{ mol } HNO_2}{47.02 \text{ g } HNO_2}\right)\left(\frac{1 \text{ mol } N_2O_3}{2 \text{ mol } HNO_2}\right)\left(\frac{22.4 \text{ L } N_2O_3}{1 \text{ mol } N_2O_3}\right) = 8.81 \text{ L } N_2O_3$$

c. 4.65×10^{23} molecules H_2O.

$$(17.3 \text{ L } N_2O_3)\left(\frac{1 \text{ mol } N_2O_3}{22.4 \text{ L } N_2O_3}\right)\left(\frac{1 \text{ mol } H_2O}{1 \text{ mol } N_2O_3}\right)\left(\frac{6.022 \times 10^{23} \text{ molecules } H_2O}{1 \text{ mol } H_2O}\right)$$
$$= 4.65 \times 10^{23} \text{ molecules } H_2O$$

7 In this problem, the provided chemical equation is already balanced, so you can proceed directly to the calculations:

a. 4.03 L Cl_2.

$$(50.0 \text{ g } PbCl_2)\left(\frac{1 \text{ mol } PbCl_2}{278.10 \text{ g } PbCl_2}\right)\left(\frac{1 \text{ mol } Cl_2}{1 \text{ mol } PbCl_2}\right)\left(\frac{22.4 \text{ L } Cl_2}{1 \text{ mol } Cl_2}\right) = 4.03 \text{ L } Cl_2$$

b. 2.97×10^{22} formula units $PbCl_4$.

$$(13.71 \text{ g } PbCl_2)\left(\frac{1 \text{ mol } PbCl_2}{278.10 \text{ g } PbCl_2}\right)\left(\frac{1 \text{ mol } PbCl_4}{1 \text{ mol } PbCl_2}\right)\left(\frac{6.022 \times 10^{23} \text{ formula units } PbCl_4}{1 \text{ mol } PbCl_4}\right)$$
$$= 2.97 \times 10^{22} \text{ formula units } PbCl_4$$

c. 67.6 g $PbCl_2$.

$$(84.8 \text{ g } PbCl_4)\left(\frac{1 \text{ mol } PbCl_4}{349.00 \text{ g } PbCl_4}\right)\left(\frac{1 \text{ mol } PbCl_2}{1 \text{ mol } PbCl_4}\right)\left(\frac{278.10 \text{ g } PbCl_2}{1 \text{ mol } PbCl_2}\right) = 67.6 \text{ g } PbCl_2$$

8 Begin limiting-reagent problems by determining which reactant is the limiting reagent. The answers to other questions build on that foundation. To find the limiting reagent, simply pick one of the reactants as a candidate. How much of your candidate reactant do you have? Use that information to calculate how much of any other reactants you'd need for a complete reaction. You can deduce the limiting and excess reagents from the results of these calculations.

a. Iron (III) oxide. In this problem, we chose iron (III) oxide as the initial candidate limiting reagent:

$$(50 \text{ g Fe}_2\text{O}_3) \left(\frac{1 \text{ mol Fe}_2\text{O}_3}{159.70 \text{ mol Fe}_2\text{O}_3} \right) \left(\frac{3 \text{ mol CO}}{1 \text{ mol Fe}_2\text{O}_3} \right) \left(\frac{28.01 \text{ g CO}}{1 \text{ mol CO}} \right) = 26 \text{ g CO}$$

Iron (III) oxide turns out to be the correct choice, because more carbon monoxide is initially present than is required to react with all the iron (III) oxide. You need 26 g of CO for every 50 g of Fe_2O_3, but you're given 67 g of CO and only 50 g of Fe_2O_3. This means you'll run out of Fe_2O_3 long before you'll run out of CO.

b. Carbon monoxide is the excess reagent, and 41 g of it will remain after the reaction is completed. Because iron (III) oxide is the limiting reagent, the excess reagent is carbon monoxide. To find the amount of excess reagent that will remain after the reaction reaches completion, first calculate how much of the excess reagent will be consumed. This calculation is identical to the one performed in part (a), so 26 g of carbon monoxide will be consumed. Next, subtract the quantity consumed from the amount originally present to obtain the amount of carbon monoxide that will remain: $67 \text{ g} - 26 \text{ g} = 41 \text{ g}$.

c. 35 g of iron; 41 g of carbon dioxide. To answer this question, do two calculations, each starting with the assumption that all of the limiting reagent is consumed:

$$(50 \text{ g Fe}_2\text{O}_3) \left(\frac{1 \text{ mol Fe}_2\text{O}_3}{159.70 \text{ mol Fe}_2\text{O}_3} \right) \left(\frac{2 \text{ mol Fe}}{1 \text{ mol Fe}_2\text{O}_3} \right) \left(\frac{55.85 \text{ g Fe}}{1 \text{ mol Fe}} \right) = 35 \text{ g Fe}$$

$$(50 \text{ g Fe}_2\text{O}_3) \left(\frac{1 \text{ mol Fe}_2\text{O}_3}{159.70 \text{ mol Fe}_2\text{O}_3} \right) \left(\frac{3 \text{ mol CO}_2}{1 \text{ mol Fe}_2\text{O}_3} \right) \left(\frac{44.01 \text{ g CO}}{1 \text{ mol CO}_2} \right) = 41 \text{ g CO}$$

9 To find the limiting reagent, simply pick one of the reactants as a candidate. Calculate how much of the other reagents you'd need to completely react with all of your available candidate reagent. Deduce the limiting and excess reagents from these calculations.

a. Sodium. In this example, we chose sodium as the initial candidate limiting reagent:

$$(25 \text{ g Na}) \left(\frac{1 \text{ mol Na}}{22.99 \text{ g Na}} \right) \left(\frac{2 \text{ mol H}_2\text{O}}{2 \text{ mol Na}} \right) \left(\frac{18.02 \text{ g H}_2\text{O}}{1 \text{ mol H}_2\text{O}} \right) = 20 \text{ g H}_2\text{O}$$

Sodium turns out to be the correct choice, because more water is initially present than is required to react with all the sodium. It takes 25 g of sodium to react with 20 g of H_2O. You're given only 25 g of sodium and 40.2 g of water, so the sodium will limit the reaction.

b. Water is the excess reagent, and 20.2 g of it will remain after the reaction is completed. Sodium is the limiting reagent, so the excess reagent is water. The calculation in part (a) reveals how much water is consumed in a complete reaction: 20 g. Because 40.2 g of water is initially present, 20.2 g of water will remain after the reaction: $40.2 \text{ g} - 20 \text{ g} = 20.2 \text{ g}$.

c. **43 g of sodium hydroxide; 12 L of hydrogen gas.** To answer this question, do two calculations, each starting with the assumption that all of the limiting reagent is consumed:

$$(25 \text{ g Na}) \left(\frac{1 \text{ mol Na}}{22.99 \text{ g Na}} \right) \left(\frac{2 \text{ mol NaOH}}{2 \text{ mol Na}} \right) \left(\frac{40.00 \text{ g NaOH}}{1 \text{ mol NaOH}} \right) = 43 \text{ g NaOH}$$

$$(25 \text{ g Na}) \left(\frac{1 \text{ mol Na}}{22.99 \text{ g Na}} \right) \left(\frac{1 \text{ mol H}_2}{2 \text{ mol Na}} \right) \left(\frac{22.4 \text{ L H}_2}{1 \text{ mol H}_2} \right) = 12 \text{ L H}_2$$

10 The first step is to identify the limiting reagent. Simply pick one of the reactants as a trial candidate for the limiting reagent, and calculate how much of the other reagents are required to react completely with the candidate.

a. **Chlorine gas.** In this example, we chose aluminum as the candidate limiting reagent:

$$(29.3 \text{ g Al}) \left(\frac{1 \text{ mol Al}}{26.98 \text{ g Al}} \right) \left(\frac{3 \text{ mol Cl}_2}{2 \text{ mol Al}} \right) \left(\frac{22.4 \text{ L Cl}_2}{1 \text{ mol Cl}_2} \right) = 36.5 \text{ L Cl}_2$$

For every 29.3 g of Al, you need 36.5 L of Cl_2 gas. More chlorine gas is required (36.5 L) to completely react with the available aluminum than is available (34.6 L), so chlorine gas is the limiting reagent.

b. **Aluminum is the excess reagent, and 1.5 g of it will remain after the reaction is completed.** To calculate how much excess reagent (aluminum, in this case) will remain after a complete reaction, first calculate how much will be consumed:

$$(34.6 \text{ L Cl}_2) \left(\frac{1 \text{ mol Cl}_2}{22.4 \text{ L Cl}_2} \right) \left(\frac{2 \text{ mol Al}}{3 \text{ mol Cl}_2} \right) \left(\frac{26.98 \text{ g Al}}{1 \text{ mol Al}} \right) = 27.8 \text{ g Al}$$

Subtract that quantity from the amount of aluminum originally present to calculate the remaining amount: 29.3 g − 27.8 g = 1.5 g.

c. **137 g of aluminum chloride.** This calculation starts with the assumption that all of the limiting reagent, chlorine gas, is consumed.

$$(34.6 \text{ L Cl}_2) \left(\frac{1 \text{ mol Cl}_2}{22.4 \text{ L Cl}_2} \right) \left(\frac{2 \text{ mol AlCl}_3}{3 \text{ mol Cl}_2} \right) \left(\frac{133.33 \text{ g AlCl}_3}{1 \text{ mol AlCl}_3} \right) = 137 \text{ g AlCl}_3$$

11 The reaction equation is already balanced, so you can proceed with the calculations.

a. **88.46% is the percent yield.** You're given an actual yield (21.61 g of sulfurous acid) and asked to calculate the percent yield. To do so, you must calculate the theoretical yield. The question clearly shows that sulfur dioxide is the limiting reagent, so begin the calculation of theoretical yield by assuming that all the sulfur dioxide is consumed:

$$(19.07 \text{ g SO}_2) \left(\frac{1 \text{ mol SO}_2}{64.07 \text{ g SO}_2} \right) \left(\frac{1 \text{ mol H}_2\text{SO}_3}{1 \text{ mol SO}_2} \right) \left(\frac{82.07 \text{ g H}_2\text{SO}_3}{1 \text{ mol H}_2\text{SO}_3} \right) = 24.43 \text{ g H}_2\text{SO}_3$$

So the theoretical yield of sulfurous acid is 24.44 g. With this information, you can calculate the percent yield:

$$\text{Percent Yield} = \left(\frac{21.61 \text{ g}}{24.43 \text{ g}} \right) \times 100 = 88.46\%$$

b. 74.6% is the percent yield. You're asked to calculate a percent yield, having been given an actual yield (27.59 g of sulfurous acid). In this case, the limiting reagent is water, so begin the calculation by assuming that all the water is consumed:

$$(8.11 \text{ g H}_2\text{O}) \left(\frac{1 \text{ mol H}_2\text{O}}{18.02 \text{ g H}_2\text{O}} \right) \left(\frac{1 \text{ mol H}_2\text{SO}_3}{1 \text{ mol H}_2\text{O}} \right) \left(\frac{82.07 \text{ g H}_2\text{SO}_3}{1 \text{ mol H}_2\text{SO}_3} \right) = 37.0 \text{ g H}_2\text{SO}_3$$

So the theoretical yield of sulfurous acid is 37.0 g. Using that information, calculate the percent yield:

$$\text{Percent Yield} = \left(\frac{27.59 \text{ g}}{37.0 \text{ g}} \right) \times 100 = 74.6\%$$

12 With a balanced equation at your disposal, you can calculate with impunity.

a. 16.1 L of nitrogen is produced. In this question, you're given the percent yield and asked to calculate an actual yield. To do so, you must know the theoretical yield. The question clearly indicates that hydrazine is the limiting reagent, so calculate the theoretical yield of nitrogen by assuming that all the hydrazine is consumed:

$$(23.4 \text{ g N}_2\text{H}_4) \left(\frac{1 \text{ mol N}_2\text{H}_4}{32.05 \text{ g N}_2\text{H}_4} \right) \left(\frac{1 \text{ mol N}_2}{1 \text{ mol N}_2\text{H}_4} \right) \left(\frac{22.4 \text{ L N}_2}{1 \text{ mol N}_2} \right) = 16.4 \text{ L N}_2$$

So the theoretical yield of nitrogen is 16.4 L. Algebraically rearrange the percent yield equation to solve for the actual yield:

$$\text{Actual Yield} = \frac{(\text{Percent Yield})(\text{Theoretical Yield})}{100}$$

Substitute for your known values and solve:

$$\text{Actual Yield} = \frac{(98)(16.4 \text{ L})}{100} = 16.1 \text{ L}$$

b. 83.3% is the percent yield. In this question, you're given the initial amounts of the reactants and an actual yield. You're asked to calculate a percent yield. To do so, you need to know the theoretical yield. The added wrinkle in this problem is that you don't initially know which reagent is limiting. As with any limiting reagent problem, pick a candidate limiting reagent. In this example, we chose hydrazine as the initial candidate:

$$(84.8 \text{ g N}_2\text{H}_4) \left(\frac{1 \text{ mol N}_2\text{H}_4}{32.05 \text{ g N}_2\text{H}_4} \right) \left(\frac{1 \text{ mol O}_2}{1 \text{ mol N}_2\text{H}_4} \right) \left(\frac{32.00 \text{ g O}_2}{1 \text{ mol O}_2} \right) = 84.7 \text{ g O}_2$$

More oxygen is required to react with the available hydrazine than is initially present, so oxygen is the limiting reagent. Knowing this, calculate a theoretical yield of water by assuming that all of the oxygen is consumed:

$$(54.7 \text{ g O}_2) \left(\frac{1 \text{ mol O}_2}{32.00 \text{ g O}_2} \right) \left(\frac{2 \text{ mol H}_2\text{O}}{1 \text{ mol O}_2} \right) \left(\frac{18.02 \text{ g H}_2\text{O}}{1 \text{ mol H}_2\text{O}} \right) = 61.6 \text{ g H}_2\text{O}$$

So the theoretical yield of water is 61.5 g. Use this information to calculate the percent yield:

$$\text{Percent Yield} = \left(\frac{51.33 \text{ g}}{61.6 \text{ g}} \right) \times 100 = 83.3\%$$

Part III
Examining Changes in Terms of Energy

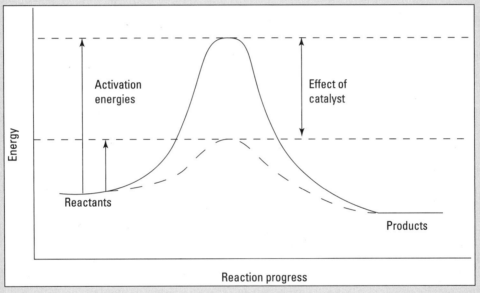

© John Wiley & Sons, Inc.

Discover how to discern differences in solid states in a free article at www.dummies.com/extras/chemistrywb.

In this part . . .

↳ Understand phases in terms of energy. There are many phases of matter, and those phases of matter depend entirely upon the behavior and energy of atoms.

↳ Get the scoop on gas laws. The behavior of compounds in the gas phase can change dramatically if you alter the pressure, temperature, or volume at which they're found. Certain formulas allow chemists to determine what type of change will take place.

↳ Survey the basics of solutions. Chemists usually measure the concentration of solutions in terms of molarity. In addition, temperature and dilution can alter solution chemistry.

↳ Beyond basic concentration measurements come colligative properties, such as freezing point depression and boiling point elevation.

↳ Take a look at thermochemistry. Different substances heat up at different rates due to differences in specific heats. Chemical reactions often give off energy, which you can calculate using Hess's law.

Chapter 10

Understanding States in Terms of Energy

· ·

· ·

*W*hen asked, children often report that solids, liquids, and gases are composed of different kinds of matter. This assumption is understandable, given the striking differences in the properties of these three states. Nevertheless, for a given type of matter at a given pressure, the fundamental difference between a solid, a liquid, and a gas actually is the *amount of energy* within the particles of matter. Understanding the states of matter (phases) in terms of energy and pressure helps to explain the different properties of those states and how matter moves between the states. We explain what you need to know in this chapter.

Describing States of Matter with the Kinetic Molecular Theory

Imagine two pool balls, each glued to either end of a spring. How many different kinds of motion could this contraption undergo? You could twist along the axis of the spring. You could bend the spring or stretch it. You could twirl the whole thing around, or you could throw it through the air. Molecules can undergo these same kinds of motions, and they do so when you supply them with energy. As collections of molecules undergo changes in energy, those collections move through the states of matter — solid, liquid, and gas. The body of ideas that explains all this is called the *kinetic molecular theory.*

Kinetic molecular theory first made a name for itself when scientists attempted to explain and predict the properties of gases and, in particular, how those properties changed with varying temperature and pressure. The idea emerged that the particles of matter within a gas (atoms or molecules) undergo a serious amount of motion as a result of the kinetic energy within them.

Kinetic energy is the energy of motion. Gas particles have a lot of kinetic energy and constantly zip about, colliding with one another or with other objects. The picture is complicated, but scientists simplified things by making several assumptions about the behavior of gas particles. These assumptions are called the *postulates* of the kinetic molecular theory. They apply to a theoretical *ideal gas:*

✔ Gases consist of tiny particles (atoms or molecules) that are always in constant random motion.

✔ Particle collisions are *elastic* (perfectly bouncy, with no loss of energy).

✔ The gas particles are assumed to neither attract nor repel one another.

✔ The average kinetic energy of a gas particle is directly proportional to the Kelvin temperature of the gas. In other words, the higher the temperature, the more kinetic energy a gas particle has.

✔ Gas particles are so small compared with the distances between them that the volume (size) of the individual particles is negligible (assumed to be zero).

The model of ideal gases explains why gas pressure increases with increased temperature. By heating a gas, you add kinetic energy to the particles; as a result, the particles collide with other objects more often and with greater force, so those objects experience greater pressure. (Check out Chapter 11 for details on gas laws.)

When atoms or molecules have less kinetic energy, or when that energy must compete with other effects (like high pressure or strong attractive forces), the matter ceases to be in the diffuse, gaseous state and comes together into one of the condensed states: liquid or solid. Here are the differences between the other two:

✔ **Liquid:** The particles within a liquid are much closer together than those in a gas. As a result, applying pressure to a liquid does very little to change the volume. The particles still have an appreciable amount of kinetic energy associated with them, so they may undergo various kinds of twisting, stretching, and vibrating motions. In addition, the particles can slide past one another fairly easily, so liquids are fluid, though less fluid than gases. Fluid matter assumes the shape of anything that contains it.

✔ **Solid:** The state of matter with the least amount of kinetic energy is the solid. In a solid, the particles are packed together quite tightly and undergo almost no movement. Therefore, solids are not fluid. Matter in the solid state may still vibrate in place or undergo other low-energy types of motion, however, depending on its temperature (in other words, on its kinetic energy).

The temperatures and pressures at which different types of matter switch between states depend on the unique properties of the atoms or molecules within that matter. Typically, particles that are very attracted to one another and have easily stackable shapes tend toward condensed states at room temperature. Particles with no mutual attraction (or that have mutual repulsion) and with not so easily stackable shapes tend toward the gaseous state. Think of a football game between fiercely rival schools. When fans of either school sit in their own section of the stands, the crowd is orderly, sitting nicely in rows. Put rival fans in the same section of the stands, however, and they'll repel each other with great energy.

Q. Are real-life gases more likely to behave like ideal gases at very high pressure or very low pressure, and why?

A. Real-life gases are more likely to behave like ideal gases when they're at very low pressure. Under very high-pressure conditions, gas particles are much closer to one another and undergo more collisions. Because of their close proximity, real-life gas particles are more likely to experience mutual attractions or repulsions. Because of their more frequent collisions, gases at very high pressure are more likely to reveal the effects of any inelasticity in their collisions (losing energy upon colliding).

1. Why does it make sense that the noble gases (which we introduce in Chapter 4) exist as gases at normal temperature and pressure?

Solve It

2. Ice floats in water. Based on the usual assumptions of kinetic molecular theory, why is this weird?

Solve It

3. At the same temperature, how would the pressure of an ideal gas differ from that of a gas with mutually attractive particles? How would it differ from that of a gas with mutually repulsive particles?

Solve It

Make a Move: Figuring Out Phase Transitions and Diagrams

 REMEMBER

Each state (solid, liquid, gas) is called a *phase.* When matter moves from one phase to another because of changes in thermal energy and/or pressure, that matter is said to undergo a *phase transition.* Moving from liquid to gas is called *boiling,* and the temperature at which boiling occurs is called the *boiling point.* Moving from solid to liquid is called *melting,* and the temperature at which melting occurs is called the *melting point.* The melting point is the same as the *freezing point,* but freezing implies matter moving from liquid to solid phase.

At the surface of a liquid, molecules can enter the gas phase more easily than elsewhere within the liquid because the motions of those molecules aren't as constrained by the molecules around them. So these surface molecules can enter the gas phase at temperatures below the liquid's characteristic boiling point. This low-temperature phase change is called *evaporation* and is very sensitive to pressure. Low pressures allow for greater evaporation, while high pressures encourage molecules to re-enter the liquid phase in a process called *condensation.*

The pressure of the gas over the surface of a liquid is called the *vapor pressure.* Understandably, liquids with low boiling points tend to have high vapor pressures because the particles are weakly attracted to each other. At the surface of a liquid, weakly interacting particles have a better chance to escape into the vapor phase, thereby increasing the vapor pressure. See how kinetic molecular theory helps make sense of things?

At the right combination of pressure and temperature, matter can move directly from a solid to a gas, or vapor. This type of phase change is called *sublimation,* and it's the kind of phase change responsible for the white mist that emanates from *dry ice,* the common name for solid carbon dioxide. Movement in the opposite direction, from gas directly into solid phase, is called *deposition.*

Table 10-1 summarizes the phase changes.

Table 10-1	Phase Changes
Phase Change (Increasing Energy)	*Phase Change (Decreasing Energy)*
Melting: Solid to liquid	Freezing: Liquid to solid
Vaporization (evaporation or boiling): Liquid to gas	Condensation: Gas to liquid
Sublimation: Solid to gas	Deposition: Gas to solid

Any given type of matter has a unique combination of pressure and temperature at the intersection of all three states. This pressure-temperature combination is called the *triple point.* At the triple point, all three phases coexist. In the case of good old H_2O, going to the triple point would produce boiling ice water. Take a moment to bask in the weirdness.

Other weird phases include plasma and supercritical fluids:

- **Plasma:** Plasma is a gas-like state in which electrons pop off gaseous atoms to produce a mixture of free electrons and *cations* (atoms or molecules with positive charge). For most types of matter, achieving the plasma state requires very high temperatures, very low pressures, or both. Matter at the surface of the sun, for example, exists as plasma.

- **Supercritical fluids:** Supercritical fluids exist under high-temperature, high-pressure conditions. For a given type of matter, there's a unique combination of temperature and pressure called the *critical point.* At temperatures and pressures higher than those at this point, the phase boundary between liquid and gas disappears, and the matter exists as a kind of liquidy gas or gassy liquid. Supercritical fluids can diffuse through solids like gases do but can also dissolve things like liquids do.

Phase diagrams are useful tools for describing the states of a given type of matter across different temperatures and pressures. A phase diagram usually displays changes in temperature on the horizontal axis and changes in pressure (in atmospheres, or atm) on the vertical axis. Lines drawn within the temperature-pressure field of the diagram represent the boundaries between phases, as shown for water and carbon dioxide in Figure 10-1.

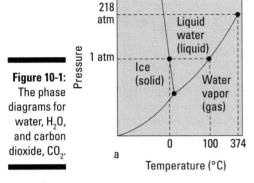

Figure 10-1:
The phase diagrams for water, H_2O, and carbon dioxide, CO_2.

© John Wiley & Sons, Inc.

Q. Ethanol (C_2H_6O) has a freezing point of –114°C. 1-propanol (C_3H_8O) has a melting point of –88°C. At 25°C (where both compounds are liquids), which one is likely to have the higher vapor pressure, and why?

A. **Ethanol** has the higher vapor pressure at 25°C. First, notice that a freezing point and a melting point are the same thing — that point is the temperature at which a substance undergoes the liquid-to-solid or solid-to-liquid phase transition. Next, compare the freezing/melting points of ethanol and propanol. Much colder temperatures must be achieved to freeze ethanol than to freeze propanol. This suggests that ethanol molecules have fewer attractive forces among themselves than propanol molecules do. At 25°C, both compounds are in liquid phase. In pure liquids whose particles have less intermolecular (between-molecule) attraction, the vapor pressure is higher because the molecules at the surface of the liquid can more easily escape into vapor (gas) phase.

4. A cup of water is put into a freezer and cools to the solid phase within an hour. The water remains at that temperature for six months. After six months, the cup is retrieved from the freezer. The cup is empty. What happened?

Solve It

5. Look at the phase diagram for carbon dioxide (CO_2) in Figure 10-1b. If you put carbon dioxide under a pressure of 4.5 atm at a temperature of 23°C, in what phase of matter would the carbon dioxide be? What are the triple point and the critical point of carbon dioxide according to the phase diagram?

Solve It

Answers to Questions on Changes of State

By this point in the chapter, you may feel depleted of kinetic energy. Hopefully, seeing whether you got some of these answers right charges you back up.

1 The noble gases are described as *noble* because due to their full valence shells, they have very low reactivity. Only very weak attractive forces occur between the atoms of a noble gas. The particles don't significantly attract one another under any but the most extreme conditions (high pressure, low temperature). Therefore, only a small amount of heat is required to push these elements into the gas phase, so they're found as gases under most conditions.

2 Kinetic molecular theory describes matter as moving from solid to liquid phase (melting) as you add energy to the sample. The added energy causes the particles to undergo greater motion and to collide with other particles more energetically. Usually, this means that a liquid is less dense than a solid sample of the same material, because the greater motion of the liquid particles prevents close packing. Solid water (ice), on the other hand, is less dense than liquid water because of the unique geometry of water crystals. Because solid water is less dense than liquid water, ice floats in water. To read more about this exception, see Chapter 22.

3 **At a given temperature, an ideal gas exerts greater pressure than a gas with mutually attractive particles, and it exerts lower pressure than a gas with mutually repulsive particles.** On average, mutual attraction allows particles to occupy a smaller volume at a given kinetic energy because the molecules are held more tightly together. Mutual repulsion causes particles to attempt to occupy a larger volume because they want to stay as far apart from each other as possible.

4 **The frozen water sublimed, moving directly from a solid to a gaseous state.** Although this process occurs slowly at the temperatures and pressures found within normal household freezers, it does occur. Try it.

5 **Carbon dioxide would be a gas. The triple point is 5.11 atm and –56.4°C. The critical point is at 73 atm and 31.1°C.** To determine the phase of the compound, plot the given temperature and pressure values on the phase diagram; if you look for where the two given measurements meet for carbon dioxide, you see they come together in the gas area of the phase diagram. To find the triple point, identify where the phase boundaries representing all three phases come together. Finally, to determine the critical point, identify where the phase boundary between liquid carbon dioxide and gaseous carbon dioxide goes away.

Chapter 11

Obeying Gas Laws

· ·

In This Chapter

▶ Seeing how pressure, volume, temperature, and moles work and play together

▶ Diffusing and effusing at different rates

· ·

At first pass, gases may seem to be the most mysterious of the states of matter. Nebulous and wispy, gases easily slip through our grip. For all their diffuse fluidity, however, gases are actually the best understood of the states. The key thing to understand about gases is that they tend to behave in the same ways — physically, if not chemically. For example, gases expand to fill the entire volume of any container in which you put them. Also, gases are easily compressed into smaller volumes. Even more so than liquids, gases easily form homogenous mixtures. Because so much open space occurs between individual gas particles, these particles are pretty laid back about the idiosyncrasies of their neighbors.

Chapter 10 sets down the basic assumptions of the kinetic molecular theory of gases, a set of ideas that explains gas properties in terms of the motions of gas particles. In summary, kinetic molecular theory describes the properties of *ideal gases,* ones that conform to the following criteria:

✔ Ideal gas particles have a volume that is insignificant compared to the volume the gas occupies as a whole. The relatively small volume of a 20-ounce soda bottle, for example, completely dwarfs the individual gas particles inside the bottle, making their sizes irrelevant to any ideal gas calculation.

✔ Ideal gases consist of very large numbers of particles in constant random motion. These particles collide with the walls of their container, and these collisions with the walls cause the pressure exerted by the gas.

✔ Ideal gas particles are neither attracted to one another nor repelled by one another.

✔ Ideal gas particles exchange energy only by means of perfectly elastic collisions — collisions in which the total kinetic energy of the particles remains constant.

✔ The average kinetic energy of ideal gas particles is directly proportional to the Kelvin (K) temperature of the gas.

Like all ideals (the ideal job, the ideal mate, and so on), ideal gases are entirely fictional. All gas particles have some volume. All gas particles have some degree of interparticle attraction or repulsion. No collision of gas particles is perfectly elastic. But imperfection is no reason to remain unemployed or lonely. Neither is it a reason to abandon the kinetic molecular theory of ideal gases. In this chapter, you're introduced to a wide variety of applications of kinetic theory, which come in the form of the so-called "gas laws."

Boyle's Law: Playing with Pressure and Volume

You deal with four important variables when working with ideal gases: pressure, volume, temperature, and the number of particles. Relationships among these four factors are the domain of the gas laws. Each variable is dependent upon the others, so altering one can change all the others as well.

The first of these relationships that was formulated into a law concerns pressure and volume. Robert Boyle, an Irish gentleman regarded by some as the first chemist (or "chymist," as his friends might have said), is typically given credit for noticing that gas pressure and volume have an inverse relationship:

$$\text{Volume} = \frac{1}{\text{Pressure}} \times \text{Constant}$$

This statement relating pressure and volume is true when the other two factors, temperature and number of particles, are fixed. Another way to express the same idea is to say that although pressure and volume may change, they do so in a way such that their product remains constant. So as a gas undergoes change in pressure (P) and volume (V) between two states, the following is true:

$$P_1 V_1 = P_2 V_2$$

The relationship between pressure and volume makes good sense in light of kinetic molecular theory. At a given temperature and number of particles, more collisions will occur at smaller volumes. These increased collisions produce greater pressure. And vice versa. Boyle had some dubious ideas about alchemy, among other things, but he really struck gold with the pressure-volume relationship in gases.

In short, pressure and volume have an indirect relationship. As one value increases, the other decreases. You can prove this idea numerically.

Q. A sealed plastic bag is filled with 1 L of air at standard temperature and pressure (STP); the values of STP are 273 K and 101.325 kilopascals (kPa). You accidentally sit on the bag. The maximum pressure the bag can withhold before popping is 500 kPa. What is the internal volume of the bag at the instant of popping?

A. **0.2 L.** The problem tells you that the bag has an initial volume, V_1, of 1 L and an initial pressure, P_1, of 101.325 kPa (the pressure at STP). The temperature doesn't matter because it remains constant. The pressure inside the bag reaches 500 kPa before popping, so that value represents P_2. You can also leave your pressure units in kilopascals, as opposed to converting them to atmospheres (atm), because both P_1 and P_2 are in the same unit (kPa). You need to convert units only when the units on the left and right side of the equation don't agree. So the only missing variable in this problem is the final volume. Solve for the final volume, V_2, by rearranging Boyle's law and then plugging in the known values:

$$V_2 = \frac{P_1 V_1}{P_2}$$

$$V_2 = \frac{(1 \text{ L})(101.325 \text{ kPa})}{(500 \text{ kPa})} = 0.2 \text{ L}$$

1. An amateur entomologist captures a particularly excellent ladybug specimen in a plastic jar. The internal volume of the jar is 0.5 L, and the air within the jar is initially at 1 atm. The bug-lover is so excited by the catch that he squeezes the jar fervently in his sweaty palm, compressing it such that the final pressure within the jar is 1.25 atm. What is the final volume of the ladybug's prison?

Solve It

2. A container has an internal volume of 3 L. This volume is divided equally in two by a gas-tight seal. On one half of the seal, neon gas resides at 5 atm (think about what the volume of this portion of the gas is). The other half of the container is kept under vacuum. Suddenly the internal seal is broken and the gas from the first chamber expands to fill the entire container. What is the final pressure within the container?

Solve It

Charles's Law and Absolute Zero: Looking at Volume and Temperature

Lest the Irish have all the gassy fun, the French contributed a gas law of their own. History attributes the following law to French chemist Jacques Charles. Charles discovered a direct, linear relationship between the volume and the temperature of a gas:

$$Volume = Temperature \times Constant$$

 This statement is true when the other two factors, pressure and number of particles, are fixed. Another way to express the same idea is to say that although temperature and volume may change, they do so in such a way that their ratio remains constant. More simply put, temperature and volume have a direct relationship; if you increase one variable, the other will increase as well. So as a gas undergoes change in temperature (T) and volume (V) between two states, the following is true:

$$\frac{V_1}{T_1} = \frac{V_2}{T_2}$$

Not to be outdone by the French, another Irish scientist took Charles's observations and ran with them. William Thomson, eventually to be known as Lord Kelvin, took stock of all the data available in his mid-19th century heyday and noticed a few things:

- ✔ Plotting the volume of a gas versus its temperature always produced a straight line.

- ✔ Extending these various lines caused them all to converge at a single point, corresponding to a single temperature at zero volume. This temperature — though not directly accessible in experiments — was about –273°C. Kelvin took the opportunity to enshrine himself in the annals of scientific history by declaring that temperature to be *absolute zero,* the lowest temperature possible.

This declaration had at least two immediate benefits. First, it happened to be correct. Second, it allowed Kelvin to create the Kelvin temperature scale, with absolute zero as the Official Zero. Using the Kelvin scale (where $°C = K – 273$), everything makes a whole lot more sense. For example, doubling the Kelvin temperature of a gas doubles the volume of that gas.

When you work with Charles's law, converting Celsius temperatures to the Kelvin scale is crucial. If you don't make the conversion, your answer will be incorrect.

Q. A red rubber dodge ball, which sits in a 20.0°C basement, is filled with 3.50 L of compressed air. Eager to begin practice for the impending dodge ball season, Vince reclaims the ball and takes it outside. After a few hours of practice, the well-sealed ball has a volume of 3.00 L. What's the temperature outside in degrees Celsius?

A. **–22°C.** The question provides an initial temperature, an initial volume, and a final volume. You're asked to find the final temperature, T_2. Apply Charles's law, plugging in the known values and solving for the final temperature. But take care — Charles's law requires you to convert all temperatures to kelvins (where

$K = °C + 273$). The correct temperature to use here is 293 K (20.0°C + 273). After identifying your variables, plug your numbers into the rearranged equation and solve:

$$\frac{V_1}{T_1} = \frac{V_2}{T_2}$$

$$T_2 = \frac{T_1 V_2}{V_1}$$

$$= \frac{(293 \text{ K})(3.00 \text{ L})}{3.50 \text{ L}} = 251 \text{ K}$$

After you determine your value in kelvins (251), you must convert back to degrees Celsius by subtracting 273 from your answer. Thus, you get –22°C.

3. A small hot air balloon has an initial volume of 300 L. To move higher into the air, the pilot runs the burner for a long period of time. The long burn increases the temperature of the air within the balloon from 40.0°C to 50.0°C. What is the new volume of the balloon?

Solve It

4. Always helpful, Danny persuades his little sister Suzie that her Very Special Birthday Balloon will last much longer if she puts it in the freezer in the basement for a while. The temperature in the house is 20.0°C. The balloon has an initial volume of 0.250 L. If the balloon has collapsed to 0.200 L by the time Suzie catches on to Danny's devious deed, what is the temperature inside the freezer in degrees Celsius?

Solve It

The Combined and Ideal Gas Laws: Working with Pressure, Volume, and Temperature

 Boyle's and Charles's laws are convenient if you happen to find yourself in situations where only two factors change at a time. But the universe is rarely so well-behaved. What if pressure, temperature, and volume all change at the same time? Are aspirin and a nap the only solution? No. Enter the *combined gas law:*

$$\frac{P_1 V_1}{T_1} = \frac{P_2 V_2}{T_2}$$

Of course, the real universe can fight back by changing another variable. In the real universe, for example, tires spring leaks. In such a situation, gas particles escape the confines of the tire. This escape decreases the number of particles, *n,* within the tire. Cranky, tire-iron wielding motorists on the side of the road will attest that decreasing *n* decreases volume. This relationship is sometimes expressed as *Avogadro's law:*

Volume = Number of particles × Constant

The final word on ideal gas behavior summarizes all four variables (pressure, temperature, volume, and number of particles) in one easy-to-use equation called the *ideal gas law:*

$$PV = nRT$$

Here, R is the *gas constant,* the one quantity of the equation that can't change. Of course, the exact identity of this constant depends on the units you're using for pressure, temperature, and volume. The most common form of the gas constant that chemists use is $R = 0.08206$ L·atm/(mol·K). Quite a few units are present, but each of them serves a purpose and must be used when solving problems with R. If the parentheses confuse you, you can also write the value of R like this:

$$R = 0.08206 \frac{\text{L} \cdot \text{atm}}{\text{mol} \cdot \text{K}}$$

Alternatively, you may encounter $R = 8.314$ L·kPa/(mol·K).

The key to solving problems with the ideal gas law and the combined gas law is starting with the correct equation in the first place. Throughout this chapter, you see multiple equations containing P, V, and T. To avoid confusion, identify all the variables given in a problem and see which equation uses those variables. For example, if the problem mentions n (moles), you have to use the ideal gas law, which is the only equation with the variable n in it. If you're given P and V in a problem but temperature and moles aren't mentioned at all, then you're going to use Boyle's law because it's the only equation that contains only P and V. Make sure you identify your equation correctly to begin with, and you'll be on the right path!

As a final gift, you may like the following equations. Sometimes working with the combined gas law can be a serious pain in the butt due to all the algebra involved with rearranging the equations endlessly. So here you go: the combined gas law solved for every variable! Please use this amazing new information sparingly.

	Initial	Final
Pressure	$P_1 = \dfrac{P_2 V_2 T_1}{T_2 V_1}$	$P_2 = \dfrac{P_1 V_1 T_2}{T_1 V_2}$
Volume	$V_1 = \dfrac{T_1 P_2 V_2}{T_2 P_1}$	$V_2 = \dfrac{T_2 P_1 V_1}{T_1 P_2}$
Temperature	$T_1 = \dfrac{P_1 V_1 T_2}{P_2 V_2}$	$T_2 = \dfrac{P_2 V_2 T_1}{P_1 V_1}$

Q. A 0.80 L container holds 10 mol of helium. The temperature of the container is 10°C. What's the internal pressure of the container?

A. **290 atm.** Consider your known and unknown variables. You're given volume, moles (number of particles), and temperature. You're asked to calculate pressure. The equation that fills the bill is the ideal gas law, $PV = nRT$. Rearrange the equation to solve for P so that $P = (nRT)/V$.

One key to using R in the ideal gas law is ensuring that all your units in the problem match with the units for R. In this problem, you have values in degrees Celsius, liters, and moles. You must convert the 10°C to the Kelvin scale by adding 273 to the temperature, giving you 283 K. In addition, you can determine that your answer for pressure will have the unit atmospheres (atm), the pressure unit given in R.

Next, plug in your known values and solve:

$$P = \frac{(10 \text{ mol}) \left(0.08206 \ \frac{\text{L·atm}}{\text{mol·K}}\right) (283 \text{ K})}{0.80 \text{ L}} = 290 \text{ atm}$$

That's nearly 300 times normal atmospheric pressure. Stay away from that container!

---

5. The 0.80 L container from the example question breaks a seal. The pressure inside the container is 290 atm, and the temperature is 283 K. Because the container holds a poisonous gas, the container was stored inside a larger, vacuum-sealed container. After the poisonous gas expands to fill the newly available volume, the gas reaches STP (273 K and 1 atm). What is the total volume of the secondary container?

Solve It

6. A container with a volume of 15.0 L contains oxygen. The gas is at a temperature of 29.0°C and a pressure of 98.69 atm. How many moles of gas occupy that container?

Solve It

7. The volume of a whoopee cushion is 0.450 L at 27.0°C and 105 kPa. Danny has placed one such practical joke device on the chair of his unsuspecting Aunt Bertha. Unbeknownst to Danny, this particular whoopee cushion suffers from a construction defect that sometimes blocks normal outgassing and ruins the flatulence effect. So even when the cushion receives the full force of Aunt Bertha's ample behind, the blockage prevents deflation. The cushion sustains the pressure exerted by Bertha so that the internal pressure becomes 200 kPa. As she sits on the cushion, Bertha warms its contents a full 10.0°C. At last, and to Danny's profound satisfaction, the cushion explodes. What volume of air does it expel?

Solve It

Mixing It Up with Dalton's Law of Partial Pressures

Gases mix. Gases mix better than liquids do and infinitely better than solids. So what's the relationship between the total pressure of a gaseous mixture and the pressure contributions of the individual gases? Here's a satisfyingly simple answer: Each individual gas within the mixture contributes a partial pressure, and adding the partial pressures yields the total pressure. This relationship is summarized by *Dalton's law of partial pressures* for a mixture of individual gases:

$$P_{total} = P_1 + P_2 + P_3 + \ldots + P_n$$

This relationship makes sense if you think about pressure in terms of kinetic molecular theory. Adding a gaseous sample into a particular volume that already contains other gases increases the number of particles in that space. Because pressure depends on the number of particles colliding with the container walls, increasing the number of particles increases the pressure proportionally.

There's no one specific pressure unit you have to use when doing problems with Dalton's partial-pressures equation. As long as the pressure units for all the gases are the same, you're good to go. However, if all the pressures given aren't in the same units, then some conversion must take place!

Q. A chemist designs an experiment to study the chemistry of the atmosphere of the early Earth. She constructs an apparatus to combine pure samples of the primary volcanic gases that made up the atmosphere billions of years ago: carbon dioxide, ammonia, and water vapor. If the partial pressures of these gases are 50 kPa, 80 kPa, and 120 kPa, respectively, what's the pressure of the resulting mixture?

A. **250 kPa.** However difficult early-Earth atmospheric chemistry may prove to be, this particular problem is a simple one. Dalton's law states that the total pressure is simply the sum of the partial pressures of the component gases:

$$P_{total} = P_{CO_2} + P_{NH_3} + P_{H_2O}$$
$$= 50 \text{ kPa} + 80 \text{ kPa} + 120 \text{ kPa}$$
$$= 250 \text{ kPa}$$

---

8. A chemist adds solid zinc powder to a solution of hydrochloric acid to initiate the following reaction:

$$\text{Zn } (s) + 2 \text{ HCl } (aq) \rightarrow \text{ZnCl}_2(aq) + \text{H}_2(g)$$

The chemist inverts a test tube and immerses the open mouth into the reaction beaker to collect the hydrogen gas that bubbles up from the solution. The reaction proceeds to equilibrium. At the end of the experiment, the water levels within the tube and outside the tube are equal. The pressure in the lab is 101.325 kPa, and the temperature of all components is 298 K. The vapor pressure of water at 298 K is 3.17 kPa. What is the partial pressure of hydrogen gas trapped in the tube?

Solve It

Diffusing and Effusing with Graham's Law

"Wake up and smell the coffee." This command is usually issued in a scornful tone, but most people who've awakened to the smell of coffee remember the event fondly. The morning gift of coffee aroma is made possible by a phenomenon called *diffusion*. Diffusion is the movement of a substance from an area of higher concentration to an area of lower concentration. Diffusion occurs spontaneously, on its own. Diffusion leads to mixing, eventually producing a homogenous mixture in which the concentration of any gaseous component is equal throughout an entire volume. Of course, that state of complete diffusion is an equilibrium state, and achieving equilibrium can take time.

Different gases diffuse at different rates, depending on their molar masses (see Chapter 7 for details on molar masses). You can compare the rates at which two gases diffuse using *Graham's law.* Graham's law also applies to *effusion,* the process in which gas molecules flow through a small hole in a container. Whether gases diffuse or effuse, they do so at a rate inversely proportional to the square root of their molar mass. In other words, more massive gas molecules diffuse and effuse more slowly than less massive gas molecules. So for Gases A and B, the following applies:

$$\frac{\text{Rate A}}{\text{Rate B}} = \sqrt{\left(\frac{\text{Molar mass B}}{\text{Molar mass A}}\right)}$$

Note: For the problems in this book, round your molar masses to two decimal places before plugging them into the formula.

By far, the most important part of solving effusion problems is identifying which gas you'll identify as Gas A and which gas you'll identify as Gas B when you plug your values into the equation. Don't switch them up!

Q. How much faster does hydrogen gas effuse than neon gas?

A. **3.16 times faster.** *Hydrogen gas* refers to H_2 because hydrogen is a diatomic element. Consult your periodic table (or your memory, if you're that good) to obtain the molar masses of hydrogen gas (2.02 g/mol) and neon gas (20.18 g/mol). Finally, plug those values into the appropriate places within Graham's law, and you can see the ratio of effusion speed. In this example, we chose hydrogen as Gas A and neon as Gas B.

$$\frac{\text{Rate } H_2}{\text{Rate Ne}} = \sqrt{\left(\frac{20.18 \text{ g/mol}}{2.02 \text{ g/mol}}\right)} = \sqrt{9.99}$$

The answer you get to this problem is 3.16. Putting this number over 1 can help you understand your answer. The ratio 3.16/1 means is that for every 3.16 mol of hydrogen gas that effuses, 1.00 mol of neon gas will effuse. This ratio is designed to compare rates. So hydrogen gas effuses 3.16 times faster than neon.

---

9. Mystery Gas A effuses 4.0 times faster than oxygen. What is the likely identity of the mystery gas?

Solve It

Answers to Questions on Gas Laws

You've answered the practice questions on gas behavior. Were your answers ideal? Check them here. No pressure.

1 **0.4 L.** You're given an initial pressure, an initial volume, and a final pressure. Boyle's law leaves you with one unknown: final volume. Solve for the final volume by plugging in the known values:

$$P_1 V_1 = P_2 V_2$$
$$V_2 = \frac{P_1 V_1}{P_2}$$
$$V_2 = \frac{(1 \text{ atm})(0.5 \text{ L})}{1.25 \text{ atm}}$$
$$V_2 = 0.4 \text{ L}$$

2 **2.5 atm.** Under the initial conditions, gas at 5 atm resides in a 1.5 L volume (half of the container's 3 L internal volume). When the seal is removed, the entire 3 L of the container becomes available to the gas, which expands to occupy the new volume. Predictably, its pressure decreases. To calculate the new pressure, P_2, plug in the known values and solve:

$$P_1 V_1 = P_2 V_2$$
$$P_2 = \frac{P_1 V_1}{V_2}$$
$$P_2 = \frac{(5 \text{ atm})(1.5 \text{ L})}{3 \text{ L}}$$
$$P_2 = 2.5 \text{ atm}$$

3 **310 L.** You're given the initial volume and initial temperature of the balloon as well as the balloon's final temperature. Apply Charles's law, plugging in the known values and solving for the final volume. Be careful — you need to express all temperatures in units of kelvins by adding 273 to Celsius temperatures, so your initial temperature is 313 K (40.0°C + 273), and your final temperature is 323 K (50.0°C + 273):

$$\frac{V_1}{T_1} = \frac{V_2}{T_2}$$
$$V_2 = \frac{V_1 T_2}{T_1}$$
$$V_2 = \frac{(300 \text{ L})(323 \text{ K})}{313 \text{ K}}$$
$$V_2 = 310 \text{ L}$$

4 **–38.6°C.** Charles's law is the method here. The unknown is the final temperature, T_2. You're given an initial temperature as well as the initial and final volumes. After converting the initial temperature to units of kelvins ($20.0°C + 273 = 293$ K), plug in the known values and solve for final temperature:

$$\frac{V_1}{T_1} = \frac{V_2}{T_2}$$

$$T_2 = \frac{T_1 V_2}{V_1}$$

$$T_2 = \frac{(293 \text{ K})(0.200 \text{ L})}{0.250 \text{ L}}$$

$$T_2 = 234.4 \text{ K}$$

The question asks for the answer in degrees Celsius, so convert the temperature: $234.4 \text{ K} - 273 = -38.6°C$.

Pay attention to little details like final units when you're answering chemistry questions. They're easy to overlook, but they make all the difference between getting the answer right and getting it wrong.

5 **224 L.** The number of moles of gas (10 mol) remains constant. The other three factors (pressure, temperature, and volume) all change between initial and final states, so you need to use the combined gas law. The initial values (290 atm, 283 K, 0.80 L) all come from the example problem. The final temperature and pressure are known (273 K, 1 atm) because the question states that the gas ends up at STP. So the only unknown is the final volume. Rearrange the combined gas law to solve for this value:

$$\frac{P_1 V_1}{T_1} = \frac{P_2 V_2}{T_2}$$

$$V_2 = \frac{P_1 V_1 T_2}{T_1 P_2}$$

$$V_2 = \frac{(290 \text{ atm})(0.80 \text{ L})(273 \text{ K})}{(283 \text{ K})(1 \text{ atm})}$$

$$V_2 = 224 \text{ L}$$

6 **59.7 mol.** This problem simply requires the ideal gas law, arranged to solve for number of moles, n. The ideal gas law involves the constant R, so your units must match the units of the constant. You have atmospheres, liters, and degrees Celsius, so you must convert the Celsius temperature to the Kelvin scale by adding 273 to it: $29.0°C + 273 = 302$ K. Plug in all the values and solve:

$$PV = nRT$$

$$n = \frac{PV}{RT}$$

$$n = \frac{(98.69 \text{ atm})(15.0 \text{ L})}{\left(0.08206 \, \frac{\text{L} \cdot \text{atm}}{\text{mol} \cdot \text{K}}\right)(302 \text{ K})}$$

$$n = 59.7 \text{ mol}$$

7 **0.244 L.** You're given an initial volume, initial temperature, and initial pressure. You're also given a final pressure. The problem doesn't specifically tell you what the final temperature is, but you're told the temperature increases by 10°C, so you can determine the final temperatire by adding 10 to the initial temperature. (Be sure to then convert both temperatures to kelvins. The initial temperature is 27.0°C+273=300 K, and the final temperature is 37°C+273=310 K.) The only unknown is final volume. Rearrange the combined gas law to solve for final volume, V_2:

$$\frac{P_1 V_1}{T_1} = \frac{P_2 V_2}{T_2}$$

$$V_2 = \frac{P_1 V_1 T_2}{T_1 P_2}$$

$$V_2 = \frac{(105 \text{ kPa})(0.450 \text{ L})(310 \text{ K})}{(300 \text{ K})(200 \text{ kPa})}$$

$$V_2 = 0.244 \text{ L}$$

Even though the final temperature is higher than the initial temperature, the final volume is much smaller than the initial volume. In effect, 10 trifling degrees is no match for the pressure exerted by Bertha's posterior.

8 **98.1 kPa.** The system has come to equilibrium, so the interior of the tube contains a gaseous mixture of hydrogen gas and water vapor. Because the water levels inside and outside the tube are equal, you know that the total pressure inside the tube equals the ambient pressure of the lab, 101.325 kPa. The total pressure includes the partial pressure contributions from hydrogen gas and from water vapor. Set up an equation using Dalton's law, rearrange the equation to solve for the pressure of just the hydrogen gas, plug in your numbers, and solve:

$$P_{total} = P_{H_2} + P_{H_2O}$$

$$P_{H_2} = P_{total} - P_{H_2O}$$

$$P_{H_2} = 101.325 \text{ kPa} - 3.17 \text{ kPa}$$

$$= 98.1 \text{ kPa}$$

9 **Hydrogen, H_2.** The question states that the ratio of the rates is 4.0. Oxygen gas is a diatomic element, so it's written as O_2 and has a molar mass of 32.00 g/mol. Substitute these known values into Graham's law to determine the molar mass of the unknown gas. The problem states that the unknown gas effuses at a rate 4.0 times faster than oxygen, so put the unknown gas over oxygen for the ratio. (In short, the unknown gas is A, and the oxygen is B).

$$\frac{\text{Rate A}}{\text{Rate } O_2} = \sqrt{\frac{\text{Molar mass } O_2}{\text{Molar mass A}}}$$

$$4.0 = \sqrt{\frac{32.00 \text{ g/mol}}{\text{Molar mass A}}}$$

Square both sides of this equation to cancel out the square root and then solve for the molar mass of unknown Gas A:

$$16.00 = \frac{32.00 \text{ g/mol}}{\text{Molar mass A}}$$

$$\text{Molar mass A} = \frac{32.00 \text{ g/mol}}{16.00}$$

$$= 2.00 \text{ g/mol}$$

After you have the molar mass, you can check the periodic table to determine the identity of the element by matching up the molar mass. No element specifically matches the 2.00 g/mol; however, if you remember that hydrogen is diatomic and is always written as H_2, you can identify the unknown gas as hydrogen!

Chapter 12

Dissolving into Solutions

Compounds can form mixtures. When compounds mix completely, right down to the level of individual molecules, we call the mixture a *solution*. Each type of compound in a solution is called a *component*. The component present in the largest amount is usually called the *solvent*. The other components are called *solutes*. Although most people think "liquid" when they think of solutions, a solution can be a solid, liquid, or gas. The only criterion is that the components are completely intermixed. We explain what you need to know about solutions in this chapter.

Seeing Different Forces at Work in Solubility

For gases, forming a solution is a straightforward process. Gases simply diffuse into a common volume (see Chapter 11 for more about diffusion). Things are a bit more complicated for condensed states like liquids and solids. In liquids and solids, molecules or ions are crammed so closely together that *intermolecular forces* are very important. Examples of these kinds of forces include ion-dipole, dipole-dipole, hydrogen bonding, and London (dispersion) forces. We touch on the physical underpinnings of these forces in Chapter 5.

Introducing a solute into a solvent initiates a tournament of forces. Attractive forces between solute and solvent compete with attractive solute-solute and solvent-solvent forces. A solution forms only to the extent that solute-solvent forces dominate over the others. The process in which solvent molecules compete and win in the tournament of forces is called *solvation* or, in the specific case where water is the solvent, *hydration*. Solvated solutes are surrounded by solvent molecules. When solute ions or molecules become separated from one another and surrounded in this way, we say they're *dissolved*.

Imagine that the members of a ridiculously popular boy band exit their hotel to be greeted by an assembled throng of fans and the media. The band members attempt to cling to each other but are soon overwhelmed by the crowd's ceaseless, repeated attempts to get closer. Soon, each member of the band is surrounded by his own attending shell of reporters and hyperventilating teenage girls. That boy band was just dissolved much like solutes are dissolved by solvents.

The tournament of forces plays out differently among different combinations of components. In mixtures where solute and solvent are strongly attracted to one another, more solute can be dissolved. One factor that always tends to favor dissolution is *entropy,* a kind of disorder or "randomness" within a system. Dissolved solutes are less ordered than undissolved solutes. There are three major types of solutions that you'll see in a general chemistry class:

- ✔ **Saturated:** Beyond a certain point, adding more solute to a solution doesn't result in a greater amount of solvation. At this point, the solution is in dynamic equilibrium; the rate at which solute becomes solvated equals the rate at which dissolved solute *crystallizes,* or falls out of solution. A solution in this state is *saturated.*

- ✔ **Unsaturated:** An *unsaturated* solution is one that can accommodate more solute.

- ✔ **Supersaturated:** A *supersaturated* solution is one in which more solute is dissolved than is necessary to make a saturated solution. A supersaturated solution is unstable; solute molecules may crash out of solution given the slightest perturbation. The situation is like that of Wile E. Coyote, who runs off a cliff and remains suspended in the air until he looks down — at which point he inevitably falls. Supersaturated solutions are most easily created by heating the solution to the point where it can accommodate more solute than it could normally handle.

The concentration of solute required to make a saturated solution is the *solubility* of that solute. Solubility varies with the conditions of the solution. The same solute may have different solubility in different solvents, at different temperatures, and so on.

When one liquid is added to another, the extent to which they intermix is called *miscibility.* Typically, liquids that have similar properties mix well — they're *miscible.* Liquids with dissimilar properties often don't mix well — they're *immiscible.* You can summarize this pattern with the phrase "like dissolves like."

Alternatively, you may understand miscibility in terms of the Italian Salad Dressing Principle. Inspect a bottle of Italian salad dressing that has been sitting in your refrigerator. The dressing consists of two distinct layers, an oily layer and a watery layer. Before using, you must shake the bottle to temporarily mix the layers. Eventually, they'll separate again because water is polar and oil is nonpolar. (See Chapter 5 if the distinction between polar and nonpolar is lost on you.) Polar and nonpolar liquids mix poorly, though occasionally with delicious consequences.

Comparing polarity between components is often a good way to predict solubility, regardless of whether those components are liquid, solid, or gas. Why is polarity such a good predictor? Because polarity is central to the tournament of forces that underlies solubility. So solids held together by ionic bonds (the most polar type of bond) or polar covalent bonds tend to dissolve well in polar solvents, like water.

Q. Sodium chloride dissolves more than 25 times better in water than in methanol. Explain this difference, referring to the structure and properties of water, methanol, and sodium chloride.

A. Sodium chloride (NaCl) is an ionic solid, a lattice composed of sodium cations (atoms with positive charge) alternating with chlorine anions (atoms with negative charge). A *lattice* has a highly regular, idealized geometry and is held together by ionic bonds, the most polar type of bond. To dissolve NaCl, a solvent must be able to engage in very polar interactions with these ions and do so with near-ideal geometry. The structure and properties of water (which is polar) are better suited to this task than are those of methanol (see the following figure). The two O–H bonds of water (on the left) partially sum to produce a

strong dipole along the mirror-image plane of the molecule that runs between the two hydrogen atoms. Methanol (on the right) is also polar, due largely to its own O–H bond, but is less polar than water. In solution, water molecules can orient their dipoles cleanly and in either of two directions to interact favorably with Na+ or Cl– ions. Methanol molecules can engage in favorable interactions with these ions, too, but not nearly as well as water.

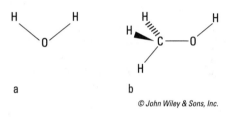

a   b

© John Wiley & Sons, Inc.

1. *Lattice energy* is a measure of the strength of the interactions between ions in the lattice of an ionic solid, measured in kilojoules per mole (kJ/mol). The greater the lattice energy, the stronger the ion-ion interactions. Here's a table of ionic solids and their associated lattice energies. Rank these ionic solids in order of their solubility in water, starting with the most soluble.

Sodium Salt	Lattice Energy (kJ/mol)
NaBr	747
NaCl	787
NaF	923
NaI	704

Solve It

2. Ethanol, CH_3CH_2OH, is miscible with water. Octanol, $CH_3(CH_2)_7OH$, isn't miscible in water. Is sucrose ($C_{12}H_{22}O_{11}$, table sugar) likely to be more soluble in ethanol or octanol? Why?

Solve It

Concentrating on Molarity and Percent Solutions

Different solutes dissolve to different extents in different solvents in different conditions. How can anybody keep track of all these differences? Chemists do so by measuring concentration. Qualitatively, a solution with a large amount of solute is said to be *concentrated*. A solution with only a small amount of solute is said to be *dilute*. As you may suspect, simply describing a solution as concentrated or dilute is usually about as useful as calling it "pretty" or naming it "Fifi." We need numbers. Two important ways to measure concentration are *molarity* and *percent solution*.

Molarity relates the amount of solute to the volume of the solution:

$$\text{Molarity } (M) = \frac{\text{Moles of solute}}{\text{Liters of solution}}$$

To calculate molarity, you may have to use conversion factors to move between units. For example, if you're given the mass of a solute in grams, use the molar mass (usually rounded to two decimal places) of that solute to convert the given mass into moles (check out Chapter 7 for an introduction to molar mass). If you're given the volume of solution in milliliters or some other unit, you need to convert that volume into liters. Units are the very first thing to check if you get a problem wrong when using molarity. Make sure your units are correct!

The units of molarity are always moles per liter (mol/L or $\text{mol} \cdot \text{L}^{-1}$). These units are often abbreviated as M and referred to as "molar." Thus, 0.25 M KOH(aq) is described as "Point two-five molar potassium hydroxide," and it contains 0.25 mol of KOH per liter of solution. Note that this does *not* mean that there are 0.25 mol KOH per liter of *solvent* (water, in this case) — only the final volume of the solution (solute plus solvent) is important in molarity.

Like other units, the unit of molarity can be modified by standard prefixes, as in millimolar (mM, which equals 10^{-3} mol/L) and micromolar (μM, which equals 10^{-6} mol/L).

Percent solution is another common way to express concentration. The precise units of percent solution typically depend on the phase of each component. For solids dissolved in liquids, mass percent is usually used:

$$\text{Mass \%} = 100\% \times \left(\frac{\text{Mass of solute}}{\text{Total mass of solution}} \right)$$

This kind of measurement is sometimes called a *mass-mass percent solution* because one mass is divided by another. Very dilute concentrations (as in the concentration of a contaminant in drinking water) are sometimes expressed as a special mass percent called *parts per million* (ppm) or *parts per billion* (ppb). In these metrics, the mass of the solute is divided by the total mass of the solution, and the resulting fraction is multiplied by 10^6 (ppm) or by 10^9 (ppb).

Sometimes, the term *percent solution* is used to describe concentration in terms of the final volume of solution instead of the final mass. For example:

- "5% $Mg(OH)_2$," can mean 5 g magnesium hydroxide in 100 mL final volume. This is a mass-volume percent solution.

- "2% H_2O_2," can mean 2 mL hydrogen peroxide in 100 mL final volume. This is a volume-volume percent solution.

Clearly, paying attention to units is important when working with concentration. Only by observing which units are attached to a measurement can you determine whether you're working with molarity, with mass percent, or with a mass-mass, mass-volume, or volume-volume percent solution.

Q. Calculate the molarity and the mass-volume percent solution obtained by dissolving 102.9 g H_3PO_4 into 642 mL final volume of solution. Be sure to use proper units. (***Hint:*** 642 mL = 0.642 L)

A. First, calculate the molarity. Before you can use the molarity formula, though, you must convert grams of H_3PO_4 to moles (see Chapter 7 for details on how to do so):

$$\left(102.9 \text{ g } H_3PO_4\right) \left(\frac{1 \text{ mol } H_3PO_4}{98.0 \text{ g } H_3PO_4}\right) = 1.05 \text{ mol } H_3PO_4$$

$$\text{Molarity} = \left(\frac{\text{mol solute}}{\text{L solution}}\right) = \left(\frac{1.05 \text{ mol } H_3PO_4}{0.642 \text{ L}}\right) = 1.64 \text{ M } H_3PO_4$$

Next, calculate the mass-volume percent solution:

$$\left(\frac{102.9 \text{ g } H_3PO_4}{642 \text{ mL}}\right) \times 100 = 16.0\% \text{ or } \left(\frac{16.0 \text{ g } H_3PO_4}{100 \text{ mL}}\right)$$

Note that the convention in molarity is to divide moles by *liters,* but the convention in mass percent is to divide grams by *milliliters.* If you prefer to think only in terms of liters (not milliliters), then simply consider mass percent as kilograms divided by liters.

---

3. Calculate the molarity of these solutions:

a. 2.0 mol NaCl in 0.872 L of solution

b. 93 g $CuSO_4$ in 390 mL of solution

c. 22 g $NaNO_3$ in 777 mL of solution

Solve It

4. How many grams of solute are in each of these solutions?

a. 671 mL of 2.0 M NaOH

b. 299 mL of 0.85 M HCl

c. 2.74 L of 0.258 M $Ca(NO_3)_2$

Solve It

5. A 15.0 M solution of ammonia, NH_3, has a density of 0.90 g/mL. What is the mass percent of this solution?

Solve It

Changing Concentrations by Making Dilutions

Real-life chemists in real-life labs don't make every solution from scratch. Instead, they make concentrated *stock solutions* and then make *dilutions* of those stocks as necessary for a given experiment.

To make a dilution, you simply add a small quantity of a concentrated stock solution to an amount of pure solvent. The resulting solution contains the amount of solute originally taken from the stock solution but disperses that solute throughout a greater volume. Therefore, the final concentration is lower; the final solution is less concentrated and more dilute.

How do you know how much of the stock solution to use and how much of the pure solvent to use? It depends on the concentration of the stock and on the concentration and volume of the final solution you want. You can answer these kinds of pressing questions by using the dilution equation, which relates concentration *(C)* and volume *(V)* between initial and final states:

$$C_1 V_1 = C_2 V_2$$

You can use the dilution equation with any units of concentration, provided you use the same units throughout the calculation. Because molarity is such a common way to express concentration, the dilution equation is sometimes expressed in the following way, where M_1 and M_2 refer to the initial and final molarity, respectively:

$$M_1 V_1 = M_2 V_2$$

Q. How would you prepare 500 mL of 0.2 M NaOH(aq) from a stock solution of 1.5 M NaOH?

A. **Add 67 mL 1.5 M NaOH stock solution to 433 mL water.** Use the dilution equation, $M_1V_1 = M_2V_2$. The initial molarity, M_1, comes from the stock solution and is therefore 1.5 M. The final molarity is the one you want in your final solution, which is 0.200 M. The final volume is the one you want for your final solution, 500 mL, which is equivalent to 0.500 L. Using these known values, you can calculate the initial volume, V_1:

$$M_1V_1 = M_2V_2$$
$$V_1 = \frac{M_2V_2}{M_1}$$
$$V_1 = \frac{(0.200\text{ M})(0.500\text{ L})}{1.5\text{ M}} = 0.067\text{ L}$$

The calculated volume is equivalent to 67 mL. The final volume of the aqueous solution is to be 500 mL, and 67 mL of this volume comes from the stock solution. The remainder, 500 mL − 67 mL = 433 mL, comes from pure solvent (water, in this case). So to prepare the solution, add 67 mL of 1.5 M stock solution to 433 mL water. Mix and enjoy!

6. What is the final concentration in molarity of a solution prepared by diluting 2.50 mL of 3.00 M KCl(aq) up to 0.175 L final volume?

Solve It

7. A certain mass of ammonium sulfate, $(NH_4)_2SO_4$, is dissolved in water to produce 1.65 L of solution. Then 80.0 mL of this solution is diluted with 120 mL of water to produce 200 mL of 0.200 M $(NH_4)_2SO_4$. What mass of ammonium sulfate was originally dissolved?

Solve It

Altering Solubility with Temperature

Increasing temperature magnifies the effects of entropy on a system. Because the entropy of a solute is usually increased when it dissolves, increasing temperature usually increases solubility — for solid and liquid solutes, anyway. Another way to understand the effect of temperature on solubility is to think about heat as a reactant in the dissolution reaction:

Solid solute + Water + Heat → Dissolved solute

Heat is usually absorbed into solute particles when a solute dissolves. Increasing temperature corresponds to added heat. So by increasing temperature, you supply a needed reactant in the dissolution reaction. (In those rare cases where dissolution releases heat, increasing temperature can decrease solubility.)

Gaseous solutes behave differently from solid or liquid solutes with respect to temperature. Increasing the temperature tends to decrease the solubility of gas in liquid. To understand this pattern, check out the concept of vapor pressure. Increasing temperature increases vapor pressure because added heat increases the kinetic energy of the particles in solution. With added energy, these particles stand a greater chance of breaking free from the intermolecular forces that hold them in solution. A classic, real-life example of temperature's effect on gas solubility is carbonated soda. Which goes flat (loses its dissolved carbon dioxide gas) more quickly: warm soda or cold soda?

Comparing gas solubility in liquids with the concept of vapor pressure highlights another important pattern: Increasing pressure increases the solubility of a gas in liquid. Just as high pressures make it more difficult for surface-dwelling liquid molecules to escape into vapor phase, high pressures inhibit the escape of gases dissolved in solvent. *Henry's law* summarizes this relationship between pressure and gas solubility:

$$\text{Solubility} = \text{Constant} \times \text{Pressure}$$

The "constant" in Henry's law is *Henry's constant,* and its value depends on the gas, solvent, and temperature. A particularly useful form of Henry's law relates the change in solubility *(S)* that accompanies a change in pressure *(P)* between two different states:

$$\frac{S_1}{P_1} = \frac{S_2}{P_2}$$

According to this relationship, tripling the pressure triples the gas solubility, for example.

Q. Henry's constant for nitrogen gas (N_2) in water at 293 K is 0.69×10^{-3} mol/(L·atm). The partial pressure of nitrogen in air at sea level is 0.78 atm. What is the solubility of N_2 in a glass of water at 20°C sitting on a coffee table within a beach house?

A. **5.4×10^{-4} mol/L.** This problem requires the direct application of Henry's law. The glass of water is at 20°C, which is equivalent to 293 K (just add 273 to any Celsius temperature to get the Kelvin equivalent). Because the glass sits within a beach house, you can assume the glass is at sea level. So you can use the provided values for Henry's constant and the partial pressure of N_2:

$$\text{Solubility} = \left(0.69 \times 10^{-3} \frac{\text{mol}}{\text{L} \cdot \text{atm}}\right)(0.78 \text{ atm}) = 5.4 \times 10^{-4} \text{ mol/L}$$

8. A chemist prepares an aqueous solution of cesium sulfate, $Cs_2(SO_4)_3$, swirling the beaker in her gloved hand to promote dissolution. She notices something, momentarily furrows her brow, and then smiles knowingly. She nestles the beaker into a bed of crushed ice within a bucket. What did the chemist notice, why was she briefly confused, and why did she place the dissolving cesium sulfate on ice?

Solve It

9. Deep-sea divers routinely operate under pressures of multiple atmospheres. One malady these divers must be concerned with is "the bends," a dangerous condition that occurs when divers rise too quickly from the depths, resulting in the rapid release of gas from blood and tissues. Why does the bends occur?

Solve It

10. Reefus readies himself for a highly productive Sunday afternoon of football-watching by arranging bags of cheesy poofs and a six-pack of grape soda around his beanbag chair. At kickoff, Reefus cracks open his first grape soda and settles in for the long haul. Three hours later, covered in cheesy crumbs, Reefus marks the end of the fourth quarter by cracking open the last of the six-pack. The soda fizzes violently all over Reefus and the beanbag chair. What happened?

Solve It

11. The grape soda preferred by Reefus is bottled under 3.5 atm of pressure. Reefus lives on a bayou at sea level (**Hint:** 1 atm). The temperature at which the soda is bottled is the same as the temperature in Reefus's living room. Assuming that the concentration of carbon dioxide in an unopened grape soda is 0.15 mol/L, what is the concentration of carbon dioxide in an opened soda that went flat while Reefus napped after the game?

Solve It

Answers to Questions on Solutions

By this point in the chapter, your brain may feel as if it has itself dissolved. Check your answers, boiling away that confusion to reveal crystalline bits of hard-earned knowledge. In other words, make sure you know what you're doing. Solutions are critical. Really.

1 **The rank order from most to least soluble is NaI, NaBr, NaCl, NaF.** As the question indicates, the greater the lattice energy is, the stronger the forces holding the ions together. Dissolving those ions means outcompeting those forces; a solution forms when attractive solute-solvent forces dominate over others, such as solute-solute bonds. Therefore, salts with lower lattice energy are typically more soluble than those with higher lattice energy.

2 **Sugar should be more soluble in ethanol than in octanol.** Like dissolves like. Chemists know from experience that sugar dissolves well in water. Therefore, you expect sugar to dissolve best in solvents that are most similar to water. Because ethanol is more miscible with water than is octanol, you expect that ethanol has solvent properties (especially polarity) more like water than octanol does.

3 Solve these kinds of problems by using the definition of molarity and conversion factors. In parts (b) and (c), you must first convert your mass in grams to moles. To do so, you divide by the molar mass from the periodic table (flip to Chapter 7 for details). In addition, be sure you convert milliliters to liters.

a. 2.3 M NaCl

$$\text{Molarity} = \frac{\text{mol solute}}{\text{L solution}} = \frac{2.0 \text{ mol NaCl}}{0.872 \text{ L}} = 2.3 \text{ M NaCl}$$

b. 1.5 M CuSO$_4$

$$\text{Moles} = \left(93 \text{ g CuSO}_4\right)\left(\frac{1 \text{ mol CuSO}_4}{159.61 \text{ g CuSO}_4}\right) = 0.58 \text{ mol CuSO}_4$$

$$\text{Volume} = (390 \text{ mL})\left(\frac{1 \text{ L}}{1,000 \text{ mL}}\right) = 0.390 \text{ L}$$

$$\text{Molarity} = \left(\frac{0.58 \text{ mol CuSO}_4}{0.390 \text{ L}}\right) = 1.5 \text{ M CuSO}_4$$

c. 0.33 M NaNO$_3$

$$\text{Moles} = \left(22 \text{ g NaNO}_3\right)\left(\frac{1 \text{ mol NaNO}_3}{85.00 \text{ g NaNO}_3}\right) = 0.26 \text{ mol NaNO}_3$$

$$\text{Volume} = 777 \text{ mL}\left(\frac{1 \text{ L}}{1,000 \text{ mL}}\right) = 0.777 \text{ L}$$

$$\text{Molarity} = \left(\frac{0.26 \text{ mol NaNO}_3}{0.777 \text{ L}}\right) = 0.33 \text{ M NaNO}_3$$

4 Again, conversion factors are the way to approach these kinds of problems. Each problem features a certain volume of solution that contains a certain solute at a certain concentration. To begin each problem, convert your volume into liters — part (c) has already done this for you. Then rearrange the molarity formula to solve for moles:

$$\text{Molarity} = \frac{\text{Moles of solute}}{\text{Liters of solution}}$$
$$\text{Moles of solute} = (\text{Molarity})(\text{Liters of solution})$$

Plug your volume and given molarity *(M)* values into the formula and solve. Finally, take your mole value and convert it into grams by multiplying it by the molar mass of your compound from the periodic table (as we explain in Chapter 7).

a. 54 g NaOH

$$\text{Volume} = 671 \text{ mL}\left(\frac{1 \text{ L}}{1{,}000 \text{ mL}}\right) = 0.671 \text{ L}$$

$$\text{Moles NaOH} = MV = \left(2.0 \ \frac{\text{mol}}{\text{L}}\right)(0.671 \text{ L}) = 1.34 \text{ mol NaOH}$$

$$\text{Mass} = (1.34 \text{ mol NaOH})\left(\frac{40.0 \text{ g NaOH}}{1 \text{ mol NaOH}}\right) = 54 \text{ g NaOH}$$

b. 9.1 g HCl

$$\text{Volume} = (299 \text{ mL})\left(\frac{1 \text{ L}}{1{,}000 \text{ mL}}\right) = 0.299 \text{ L}$$

$$\text{Moles HCl} = MV = \left(0.85 \ \frac{\text{mol}}{\text{L}}\right)(0.299 \text{ L}) = 0.25 \text{ mol HCl}$$

$$\text{Mass} = (0.25 \text{ mol HCl})\left(\frac{36.46 \text{ g HCl}}{1 \text{ mol HCl}}\right) = 9.1 \text{ g NaOH}$$

c. 116 g Ca(NO₃)₂

$$\text{Moles Ca}(NO_3)_2 = MV = \left(0.258 \ \frac{\text{mol}}{\text{L}}\right)(2.74 \text{ L}) = 0.707 \text{ mol Ca}(NO_3)_2$$

$$\text{Mass} = \left(0.707 \text{ molCa}(NO_3)_2\right)\left(\frac{164 \text{ g Ca}(NO_3)_2}{1 \text{ mol Ca}(NO_3)_2}\right) = 116 \text{ g Ca}(NO_3)_2$$

5 **28%.** To calculate mass percent, you must know the mass of solute and the mass of solution. The molarity of the solution tells you the moles of solute per volume of solution. Starting with this information, you can convert to mass of solute by means of the gram formula mass (see Chapter 7 for details on calculating the gram formula mass):

$$\text{Moles NH}_3 = \left(15 \ \frac{\text{mol NH}_3}{\text{L}}\right)(1 \text{ L}) = 15 \text{ mol NH}_3$$

$$\text{Mass} = (15 \text{ mol NH}_3)\left(\frac{17.04 \text{ g NH}_3}{1 \text{ mol NH}_3}\right) = 256 \text{ g NH}_3$$

So each liter of 15.0 M NH_3 contains 256 g of NH_3 solute. But how much mass does each liter of solution possess? Calculate the mass of the solution by using the density. Note that the problem lists the density in units of milliliters, so be sure to convert to the proper units. To keep the calculations easy, we converted the liters of solution to milliliters to match the density units:

$$\text{Volume} = (1 \text{ L}) \left(\frac{1{,}000 \text{ mL}}{1 \text{ L}} \right) = 1{,}000 \text{ mL}$$

$$\text{Density} = \frac{\text{Mass}}{\text{Volume}}$$
$$\text{Mass} = (\text{Density})(\text{Volume})$$
$$= \left(0.90 \, \frac{\text{g}}{\text{mL}} \right) (1{,}000 \text{ mL}) = 900 \text{ g of solution}$$

So 255 g of NH_3 is in every 900 g of 15.0 M NH_3. Now you can calculate the mass percent:

$$\text{Mass \%} = 100 \times \left(\frac{256 \text{ g}}{900 \text{ g}} \right) = 28\%$$

6 **4.29×10^{-2}M.** Use the dilution equation, $M_1 V_1 = M_2 V_2$. In this problem, the initial molarity is 3.00 M, the initial volume is 2.50 mL (or 2.50×10^{-3} L), and the final volume is 0.175 L. Use these known values to calculate the final molarity, M_2:

$$M_1 V_1 = M_2 V_2$$
$$M_2 = \frac{M_1 V_1}{V_2} = \frac{(3.0 \text{ M})(0.0025 \text{ L})}{(0.175 \text{ L})} = 4.29 \times 10^{-2} \text{ M}$$

7 **$109 \text{ g} \left(NH_4 \right)_2 SO_4$.** First, use the dilution equation to find the concentration of the original solution. The initial volume of the solution is given to you as 80 mL and must be converted to liters to become 0.080 L. The final volume is 200 mL (0.2 L when converted), and the final concentration is 0.2 M $\left(NH_4 \right)_2 SO_4$. You then solve for the initial concentration:

$$M_1 V_1 = M_2 V_2$$
$$M_1 = \frac{M_2 V_2}{V_1} = \frac{(0.2 \text{ M})(0.2 \text{ L})}{(0.080 \text{ L})} = 0.5 \text{ M}$$

This calculation means that the original solution contained 0.5 mol of $\left(NH_4 \right)_2 SO_4$ per liter of solution. The question indicates that 1.65 L of this original solution was prepared, so you then take the initial concentration and use the molarity formula to solve for the mole value of $\left(NH_4 \right)_2 SO_4$. When you have the mole value, you multiply it by the molar mass of $\left(NH_4 \right)_2 SO_4$ to determine the initial mass of $\left(NH_4 \right)_2 SO_4$ dissolved.

$$(1.65 \text{ L}) \left(\frac{0.5 \text{ mol} \left(NH_4 \right)_2 SO_4}{1 \text{ L}} \right) = 0.825 \text{ mol} \left(NH_4 \right)_2 SO_4$$

$$\left(0.825 \text{ mol} \left(NH_4 \right)_2 SO_4 \right) \left(\frac{132.16 \text{ g} \left(NH_4 \right)_2 SO_4}{1 \text{ mol} \left(NH_4 \right)_2 SO_4} \right) = 109 \text{ g} \left(NH_4 \right)_2 SO_4$$

8 As the chemist swirled the beaker of dissolving cesium sulfate, the beaker was becoming noticeably warmer. This observation momentarily confused her, because it suggested that the dissolution of cesium sulfate released heat, a state of affairs opposite to that usually observed with dissolving salts. Having diagnosed the situation, she cleverly turned it to her advantage. With typical salts, increasing temperature increases solubility in water, so heating a dissolving mixture can promote dissolution. In the case of cesium sulfate, however, the reverse is true: By cooling the dissolving mixture, the chemist promoted solubility of the cesium sulfate. (Again, though, it's far more common for solubility to increase as temperature increases.)

9 Deep-sea divers are exposed to high pressures during their dives, and at high pressure, gases become more soluble in the blood and tissue fluids due to Henry's law (Solubility = Constant × Pressure). So when the divers do their thing at great depth, high concentrations of these gases dissolve in the blood. If the divers rise to the surface too quickly at the end of a dive, the solubility of these dissolved gases changes too quickly in response to the diminished pressure. This situation can lead to the formation of tiny gas bubbles in the blood and tissues, which can be deadly.

10 Nothing dramatically fizzy happened when Reefus opened the first soda because that soda was still cold from the refrigerator. As the game progressed, however, the remaining sodas warmed to room temperature as they sat beside Reefus's beanbag chair. Gases (like carbon dioxide) are less soluble in warmer liquids. So when Reefus opened the warm, fourth-quarter soda, a reservoir of undissolved gas burst forth from the can.

11 4.3×10^{-2} **mol/L.** To solve this problem, use the two-state form of Henry's law:

$$\frac{S_1}{P_1} = \frac{S_2}{P_2}$$

The initial solubility and pressure are 0.15 mol/L and 3.5 atm, respectively. The final pressure is 1.0 atm. Using these known values, solve for the final solubility:

$$\frac{S_1}{P_1} = \frac{S_2}{P_2}$$

$$S_2 = \frac{S_1 P_2}{P_1}$$

$$S_2 = \frac{\left(0.15\ \frac{\text{mol}}{\text{L}}\right)(1.0\ \text{atm})}{3.5\ \text{atm}} = 0.043\ \text{mol/L}$$

Chapter 13

Playing Hot and Cold: Colligative Properties

In This Chapter

▶ Knowing the difference between molarity and molality

▶ Working with boiling point elevation and freezing point depression

▶ Deducing molecular masses from boiling and freezing point changes

As a recently minted expert in solubility and molarity (see Chapter 12), you may be ready to write off solutions as another chemistry topic mastered, but you, as a chemist worth your salt, must be aware of one final piece to the puzzle. Collectively called the *colligative properties,* these chemically important phenomena arise from the presence of solute particles in a given mass of solvent. The presence of extra particles in a formerly pure solvent has a significant impact on some of that solvent's characteristic properties, such as freezing and boiling points. This chapter walks you through these colligative properties and their consequences, and it introduces you to a new solution property: molality. No, that's not a typo. Molality is a different way to measure concentration that allows you to solve for the key colligative properties later in this chapter.

Portioning Particles: Molality and Mole Fractions

Like the difference in their names, the practical difference between *molarity* and *molality* is subtle. Take a close look at their definitions, expressed next to one another in the following equations:

$$\text{Molarity} = \frac{\text{Moles of solute}}{\text{Liters of solution}} \qquad \text{Molality} = \frac{\text{Moles of solute}}{\text{Kilograms of solvent}}$$

The numerators in molarity and molality calculations are identical, but their denominators differ greatly. Molarity deals with liters of solution, while molality deals with kilograms of solvent. A *solution* is a mixture of solvent and solute; a *solvent* is the medium into which the solute is mixed.

A further complication to the molarity/molality confusion is how to distinguish between their variables and units. The letter *m* turns out to be overused in chemistry. Instead of picking another variable (or perhaps a less confusing name that started with a nice uncommon letter like *z*), chemists decided to give molality the lowercase "script" *m* as its variable. To help you avoid uttering any four-letter words when confronted with this plethora of *m*-words and their abbreviations, we've provided Table 13-1.

Table 13-1	M Words Related to Concentration	
Name	*Variable*	*Unit Abbreviation*
Molarity	*M*	M
Molality	*m*	m
Moles		mol

Occasionally, you may be asked to calculate the *mole fraction* of a solution, which is the ratio of the number of moles of *either* solute or solvent in a solution to the total number of moles of solute *and* solvent in the solution. By the time chemists defined this quantity, however, they had finally acknowledged that they had too many *m* variables, and they gave it the variable *X*. Of course, chemists still need to distinguish between the mole fractions of the solute and the solvent, which unfortunately both start with the letter *s*. To avoid further confusion, they decided to abbreviate solute and solvent as A and B, respectively, in the general formula, although in practice, the chemical formulas of the solute and solvent are usually written as subscripts in place of A and B. For example, the mole fraction of sodium chloride in a solution would be written as X_{NaCl}.

In general, the mole ratio of the solute in a solution is expressed as

$$X_A = \frac{n_A}{n_A + n_B}$$

Where n_A is the number of moles of solute and n_B is the number of moles of solvent. The mole ratio of the solvent is then

$$X_B = \frac{n_B}{n_A + n_B}$$

These mole fractions are useful because they represent the ratio of solute to solution and solvent to solution very well and give you a general understanding of how much of your solution is solute and how much is solvent.

EXAMPLE

Q. How many grams of dihydrogen sulfide (H_2S) must you add to 750 g of water to make a 0.35 m solution?

A. **8.9 g H_2S.** This problem gives you molality and the mass of a solvent and asks you to solve for the mass of solute. Because molality involves moles and not grams of solute, you first need to solve for moles of solvent, and then you use the gram formula mass of sodium chloride to solve for the number of grams of solute (see Chapter 7 for details on how to make this conversion). Before plugging the numbers into the molality equation, you must also note that the problem has given you the mass of the solvent in grams, but the formula calls for it to be in kilograms. Moving from

grams to kilograms is equivalent to moving the decimal point three places to the left (if you need a refresher on the ins and outs of unit conversions, please refer to Chapter 2). Plugging everything you know into the equation for molality gives you the following:

$$0.35 \text{ m} = \frac{X \text{ mol } H_2S}{0.750 \text{ kg } H_2O}$$

Solving for the unknown gives you 0.26 mol of H_2S in solution. You then need to multiply this mole value by the molecular mass of H_2S to determine the number of grams that need to be added:

$$(0.26 \text{ mol } H_2S)\left(\frac{34.08 \text{ g } H_2S}{1 \text{ mol } H_2S}\right) = 8.86 \text{ g } H_2S$$

1. What is the molality of a solution made by dissolving 36 g of sodium chloride (NaCl) in 6 L of water? (Remember that the density of water is 1.0 kg/L.)

Solve It

2. How many moles of potassium iodide (KI) are required to produce a 3.5 m solution if you begin with 2.7 L of water?

Solve It

3. Calculate the mole fraction of each component in a solution containing 2.75 mol of ethanol (C_2H_6O) and 6.25 mol of water.

Solve It

Too Hot to Handle: Elevating and Calculating Boiling Points

Calculating molality is no more or less difficult than calculating molarity, so you may be asking yourself, "Why all the fuss?" Is it even worth adding another quantity and another variable to memorize? Yes! Although molarity is exceptionally convenient for calculating concentrations and working out how to make dilutions in the most efficient way, molality is reserved for the calculation of several important colligative properties, including boiling point elevation. *Boiling point elevation* refers to the tendency of a solvent's boiling point to increase when an impurity (a solute) is added to it. In fact, the more solute that is added, the greater the change in the boiling point.

Boiling point elevations are directly proportional to the molality of a solution, but chemists have found that some solvents are more susceptible to this change than others. The formula for the change in the boiling point of a solution, therefore, contains a proportionality constant, abbreviated K_b, which is a property determined experimentally and must be read from a table such as Table 13-2. The formula for the boiling point elevation is

$$\Delta T_b = K_b m$$

where m is molality. Note the use of the Greek letter delta (Δ) in the formula to indicate that you're calculating a *change in* the boiling point, not the boiling point itself. You need to add this number to the boiling point of the pure solvent to get the boiling point of the solution. The units of K_b are typically given in degrees Celsius per molality.

Table 13-2	Common K_b Values	
Solvent	**K_b in °C/m**	**Boiling Point in °C**
Acetic acid	3.07	118.1
Benzene	2.53	80.1
Camphor	5.95	204.0
Carbon tetrachloride	4.95	76.7
Cyclohexane	2.79	80.7
Ethanol	1.19	78.4
Phenol	3.56	181.7
Water	0.512	100.0

Boiling point elevations are a result of the attraction between solvent and solute particles in a solution. Colligative properties such as boiling point elevation depend on only the number of particles *in solution.* Adding solute particles increases these intermolecular attractions because more particles are around to attract one another. To boil, solvent particles must therefore achieve a greater kinetic energy to overcome this extra attractive force, which translates into a higher boiling point. (See Chapter 10 for more information on kinetic energy.)

EXAMPLE

Q. What is the boiling point of a solution containing 45.2 g of menthol ($C_{10}H_{20}O$) dissolved in 350 g of acetic acid?

A. **120.6°C.** The problem asks for the boiling point of the solution, so you know that first you have to calculate the boiling point elevation. This means you need to know the molality of the solution and the K_b value of the solvent (acetic acid). Table 13-2 tells you that the K_b of acetic acid is 3.07. To calculate the molality, you must convert 45.2 g of menthol to moles:

$$(45.2 \text{ g menthol}) \left(\frac{1 \text{ mol menthol}}{156.3 \text{ g menthol}} \right) = 0.29 \text{ mol menthol}$$

You can now calculate the molality of the solution, taking care to convert grams of acetic acid to kilograms:

$$m = \frac{0.29 \text{ mol menthol}}{0.350 \text{ kg acetic acid}} = 0.83 \text{ m}$$

Now that you have molality, you can plug it and your K_b value into the formula to find the change in boiling point:

$$\Delta T_b = \left(3.07 \frac{°C}{m} \right) (0.83 \text{ m}) = 2.5°C$$

You're not quite done, because the problem asks for the boiling point of the solution, not the change in the boiling point. Luckily, the last step is just simple arithmetic. You must add your ΔT_b to the boiling point of pure acetic acid, which, according to Table 13-2, is 118.1°C. This gives you a final boiling point of 118.1°C + 2.5°C = 120.6°C for the solution.

4. What is the boiling point of a solution containing 158 g of sodium chloride (NaCl) and 1.2 kg of water? What if the same number of moles of calcium chloride ($CaCl_2$) is added to the solvent instead? Explain why there's such a great difference in the boiling point elevation.

Solve It

5. A clumsy chemist topples a bottle of indigo dye ($C_{16}H_{10}N_2O_2$) into a beaker containing 450 g of ethanol. If the boiling point of the resulting solution is 79.2°C, how many grams of dye were in the bottle?

Solve It

How Low Can You Go? Depressing and Calculating Freezing Points

The second of the important colligative properties that you can calculate from molality is *freezing point depression*. Not only does adding solute to a solvent raise its boiling point, but it also lowers its freezing point. That's why you sprinkle salt on icy sidewalks. The salt mixes with the ice and lowers its freezing point. If this new freezing point is lower than the outside temperature, the ice melts, eliminating the spectacular wipeouts so common on salt-free sidewalks. The colder it is outside, the more salt you need to melt the ice and lower the freezing point to below the ambient temperature.

Freezing point depressions, like boiling point elevations in the preceding section, are calculated using a constant of proportionality, this time abbreviated K_f. The formula therefore becomes $\Delta T_f = K_f m$, where m is molality. To calculate the new freezing point of a compound, you must *subtract* the change in freezing point from the freezing point of the pure solvent. Table 13-3 lists several common K_f values.

Table 13-3	Common K_f Values	
Solvent	*K_f in °C/m*	*Freezing Point in °C*
Acetic acid	3.90	16.6
Benzene	5.12	5.5
Camphor	37.7	179.0
Carbon tetrachloride	30.0	−23.0
Cyclohexane	20.2	6.4
Ethanol	1.99	−114.6
Phenol	7.40	41.0
Water	1.86	0.0

Adding an impurity to a solvent makes its liquid phase more stable through the combined effects of boiling point elevation and freezing point depression. That's why you rarely see bodies of frozen salt water. The salt in the oceans lowers the freezing point of the water, making the liquid phase more stable and able to sustain temperatures slightly below 0°C.

Q. Each kilogram of seawater contains roughly 35 g of dissolved salts. Assuming that all these salts are sodium chloride, what is the freezing point of seawater?

A. **−2.23°C.** Begin by converting grams of salt to moles to figure out the molality. One mole of NaCl is equivalent to 58.4 g, so 35 g is equivalent to 0.60 mol of NaCl. You need to multiply this number by 2 to compensate for the fact that sodium chloride dissociates into twice as many particles in water, so this solution contains 1.20 mol. Next, find the molality of the solution by dividing this number of moles by the mass of the solvent (1 kg), giving a 1.20 m solution. Last, look up the K_f of water in Table 13-3 and plug all these values into the equation for freezing point depression:

$$\Delta T_f = \left(1.86\frac{°C}{m}\right)(1.20\ m) = 2.23°C$$

Because this value is merely the freezing point depression, you must subtract it from the freezing point of the pure solute to get $0°C - 2.23°C = -2.23°C$, the freezing point of seawater.

6. Antifreeze takes advantage of freezing point depression to lower the freezing point of the water in your car's engine and keep it from freezing on blistery winter drives. If antifreeze is made primarily of ethylene glycol ($C_2H_6O_2$), how much of it needs to be added to lower the freezing point of 10.0 kg of water by 15.0°C?

Solve It

7. If 15 g of silver (Ag) is dissolved into 1,500 g of ethanol (C_2H_6O), what is the freezing point of the mixture?

Solve It

Determining Molecular Masses with Boiling and Freezing Points

Just as a solid understanding of molality helps you to calculate changes in boiling and freezing points, a solid understanding of ΔT_b and ΔT_f can help you determine the molecular mass of a mystery compound that's being added to a known quantity of solvent. When you're asked to solve problems of this type, you'll always be given the mass of the mystery solute, the mass of solvent, and either the change in the freezing or boiling point or the new freezing or boiling point itself. From this information, you then follow a set of simple steps to determine the molecular mass:

1. **Find the boiling point elevation or freezing point depression.**

   If you've been given the boiling point, calculate the ΔT_b by subtracting the boiling point of the pure solvent from the number you were given. If you know the freezing point, add the freezing point of the pure solvent to it to get the ΔT_f.

2. **Look up the K_b or K_f of the solvent (refer to Tables 13-2 and 13-3).**

3. **Solve for the molality of the solution using the equation for ΔT_b or ΔT_f.**

4. **Calculate the number of moles of solute in the solution by multiplying the molality calculated in Step 3 by the given number of kilograms of solvent.**

5. Divide the given mass of solute by the number of moles calculated in Step 4. This is your molecular mass, or number of grams per mole, from which you can often guess the identity of the mystery compound.

Q. 97.30 g of a mystery compound is added to 500.0 g of water, raising its boiling point to 100.78°C. What is the molecular mass of the mystery compound?

A. **128.0 g/mol.** First subtract the boiling point of water from this new boiling point:

$$\Delta T_b = 100.78°C - 100.00°C = 0.78°C$$

Then plug this value and a K_b of 0.512 into the equation for boiling point elevation and solve for molality:

$$m = \frac{\Delta T_b}{K_b} = \frac{0.78°C}{0.512\frac{°C}{m}} = 1.52\ m$$

Next, take this molality value and multiply it by the given mass of the solvent, water, in kilograms:

$$\left(\frac{1.52\ \text{mol solute}}{1\ \text{kg H}_2\text{O}}\right)(0.5\ \text{kg H}_2\text{O}) = 0.76\ \text{mol solute}$$

Last, divide the number of grams of the mystery solute by the number of moles, giving you the molecular mass of the compound:

$$\frac{97.30\ \text{g}}{0.76\ \text{mol}} = 128.0\ \text{g/mol}$$

---

8. The freezing point of 83.2 g of carbon tetrachloride is lowered by 11.52°C when 15.0 g of a mystery compound is added to it. What is the molecular mass of this mystery compound?

Solve It

9. When 8.8 g of a mystery compound is added to 42.1 g of benzene, the boiling point is elevated to 81.9°C. What is the molecular mass of this mystery compound?

Solve It

Answers to Questions on Colligative Properties

You've tried your hand at problems on molality, boiling points, and freezing points. Are you feeling a little hot under the collar or as cool as a cucumber? Here, we present the answers to the practice problems for this chapter.

1 **0.10 m.** This problem gives you the value of your solute (sodium chloride) in grams and the value of your solvent (water) in liters. You must first convert the sodium chloride to moles. To do so, divide by the molar mass (which we discuss in Chapter 7):

$$(36 \text{ g NaCl}) \left(\frac{1 \text{ mol NaCl}}{58.44 \text{ g NaCl}} \right) = 0.62 \text{ mol NaCl}$$

Then convert liters to kilograms using the given density of water. Thankfully, this is an easy one-to-one conversion: 6 L of water becomes 6 kg of water. Finally, plug your values into the molality formula to determine the molality of your solution:

$$m = \frac{0.62 \text{ mol NaCl}}{6 \text{ kg H}_2\text{O}} = 0.10 \text{ m}$$

2 **9.5 mol.** This problem gives you your molality value and the liters of solution. You simply have to determine the number of moles of solute (potassium iodide) that you need to add to make this solution. To do so, first convert the liters of solution to kilograms of solution using the density of water (1.0 kg/L). Then plug your values into the molality formula and solve for moles of solute:

$$3.5 \text{ m} = \frac{\text{moles of solute}}{2.7 \text{ kg of solvent}} = 9.5 \text{ mol of solute}$$

3 **The solution is 31% ethanol and 69% water.** This problem tells you that the molality of ethanol $\left(n_{\text{C}_2\text{H}_6\text{O}} \right)$ is 2.75 mol and the molality of water $\left(n_{\text{H}_2\text{O}} \right)$ is 6.25 mol. Plugging these values into the equations for the mole fraction of solute and solvent yields

$$X_{\text{C}_2\text{H}_6\text{O}} = \frac{2.75 \text{ mol}}{2.75 \text{ mol} + 6.25 \text{ mol}} = 0.31$$

$$X_{\text{H}_2\text{O}} = \frac{6.25 \text{ mol}}{2.75 \text{ mol} + 6.25 \text{ mol}} = 0.69$$

4 The solution containing sodium chloride has a boiling point of **102.3°C;** the solution containing calcium chloride has a boiling point of **103.5°C.** To solve for the boiling point, you must first solve for the molality. Start by dividing 158 g NaCl by its gram formula mass (58.44 g/mol), which tells you that 2.71 mol of solute is being added to the water. Multiply this value by 2 because each molecule of NaCl splits into two particles in solution, for a total of 5.42 mol. Divide the moles by the mass of solvent (1.2 kg) to give you a molality of 4.5 m. Finally, multiply this molality by the K_b of water, 0.512°C/m, to give you a ΔT_b of 2.3°C. Add this temperature change to the boiling point of pure water (100°C) to give you a new boiling point of 102.3°C.

$$\Delta T_b = K_b m$$
$$= \left(0.512 \frac{°\text{C}}{\text{m}} \right) (4.5 \text{ m})$$
$$= 2.3°\text{C}$$

$$T_b = 100°\text{C} + 2.3°\text{C} = 102.3°\text{C}$$

If you instead add the same number of moles of calcium chloride (2.71 mol) to the water, the calcium chloride would dissociate into three particles per mole in solution. This gives you 2.71 mol×3=8.13 mol of solute in solution. As with the sodium chloride solution, divide the number of moles by the mass of solvent (1.2 kg) to get 6.8 m, and multiply by the K_b of water (0.512°C/m) to get a ΔT_b of 3.5°C. This is a difference of more than 1 degree! The difference arises because colligative properties such as boiling point elevation depend on only the number of particles *in solution*.

$$\Delta T_b = K_b m$$
$$= \left(0.512\frac{°C}{m}\right)(6.8\ m)$$
$$= 3.5°C$$

$$T_b = 100°C + 3.5°C = 103.5°C$$

5 **79 g $C_{16}H_{10}N_2O_2$.** This problem requires you to solve a boiling point elevation problem backwards. You're given the solution's boiling point, so the first thing to do is to solve for the ΔT_b of the solution. You do this by subtracting the boiling point of pure ethanol (which you find in Table 13-2) from the given boiling point of the impure solution:

$$\Delta T_b = 79.2°C - 78.4°C = 0.8°C$$

After you look up the K_b of ethanol (1.19°C/m), the only unknown in your ΔT_b equation is the molality. Solving for this gives you

$$m = \frac{\Delta T_b}{K_b} = \frac{0.8°C}{1.19\frac{°C}{m}} = 0.67\ m$$

Now that you know the molality, you can find the number of moles of solute. Be sure to convert grams of ethanol to kilograms before you plug everything into the formula.

$$0.67\ m = \frac{x\ \text{mol}\ C_{16}H_{10}N_2O_2}{0.450\ \text{kg}\ C_2H_6O}$$
$$x = 0.30\ \text{mol}\ C_{16}H_{10}N_2O_2$$

Last, translate this mole count into a mass by multiplying by the gram molecular mass of $C_{16}H_{10}N_2O_2$ (262 g/mol), giving you your final answer of 78.6 g $C_{16}H_{10}N_2O_2$.

6 **5.00 kg $C_2H_6O_2$.** Here, you've been given a freezing point depression of 15.0°C and are asked to back-solve for the number of grams of antifreeze required to make it happen. Begin by solving for the molality of the solution by plugging all the known quantities into your freezing point depression equation and solving for molality:

$$m = \frac{\Delta T_f}{K_f} = \frac{15.0°C}{1.86\frac{°C}{m}} = 8.06\ m$$

Now solve for the number of moles of the solvent ethylene glycol (antifreeze):

$$8.06\ m = \frac{x\ \text{mol}\ C_2H_6O_2}{10.0\ \text{kg}\ H_2O}$$
$$x = 80.6\ \text{mol}\ C_2H_6O_2$$

Finally, translate this mole count into a mass by multiplying by the gram molecular mass of $C_2H_6O_2$ (62.0 g/mol), giving you your answer of 4,997.2 g, which is approximately equal to 5.00 kg $C_2H_6O_2$. Lowering the freezing point of water by such a significant amount requires a solution that is one-third antifreeze by mass!

7 **–114.8°C.** First calculate the molality of the solution by converting the mass of silver into a mole count ((15 g)×(1 mol/108 g)=0.14 mol) and dividing it by the mass of ethanol in kilograms (turning 1,500 g into 1.5 kg), giving you 0.093 m. Then multiply this molality by the K_f of ethanol (1.99 °C/m), giving you a ΔT_f of 0.19°C. Last, subtract this ΔT_f value from the freezing point of pure ethanol, giving you

$$-114.6°C - 0.19°C = -114.8°C$$

8 **470 g/mol.** Carefully follow the steps outlined in "Determining Molecular Masses with Boiling and Freezing Points" to arrive at the correct answer. Step 1 is unnecessary in this case because you already have the ΔT_f, so begin by looking up the K_f of carbon tetrachloride in Table 13-3 (30.0°C/m). Plug both of these values into the equation for boiling point elevation and solve for molality:

$$m = \frac{\Delta T_f}{K_f} = \frac{11.52°C}{30.0\frac{°C}{m}} = 0.384 \text{ m}$$

Next, take this molality value and multiply it by the given value for the mass of the solvent (first converting grams to kilograms):

$$\left(\frac{0.384 \text{ mol solute}}{1 \text{ kg CCl}_4}\right)(0.0832 \text{ kg}) = 0.0319 \text{ mol solute}$$

Last, divide the number of grams of the mystery solute by the number of moles, giving you the molecular mass of the compound:

$$\left(\frac{15.0 \text{ g}}{0.0319 \text{ mol}}\right) = 470 \text{ g/mol}$$

9 **290 g/mol.** Solve for ΔT_b by subtracting the boiling point of pure benzene (refer to Table 13-2) from the given boiling point of the solution:

$$81.9°C - 80.1°C = 1.80°C$$

Plug this value and the K_b for benzene (2.53°C/m) into your ΔT_b equation and solve for the molality of the solution:

$$m = \frac{\Delta T_b}{K_b} = \frac{1.80°C}{2.53\frac{°C}{m}} = 0.71 \text{ m}$$

Next, take this molality value and multiply it by the given value for mass of solvent (first converting grams to kilograms):

$$\left(\frac{0.71 \text{ mol solute}}{1 \text{ kg benzene}}\right)(0.0421 \text{ kg}) = 0.030 \text{ mol solute}$$

Last, divide the number of grams of the mystery solute by the number of moles, giving you the molecular mass of the compound:

$$\left(\frac{8.8 \text{ g}}{0.030 \text{ mol}}\right) = 293.3 \text{g/mol}$$

The answer needs only two significant figures, so use 290 g/mol.

Chapter 14

Exploring Rates and Equilibrium

. . . .

In This Chapter

▶ Measuring reaction rates and understanding the factors that affect them

▶ Measuring equilibrium and understanding how it responds to disruption

. . . .

Most people don't like waiting. And nobody likes waiting for nothing. Researchers have tentatively concluded that chemists are people, too. It follows that chemists don't like to wait, and if they must wait, they'd prefer to get something for their trouble.

To address these concerns, chemists study things like rates and equilibrium:

✔ Rates tell chemists how long they'll have to wait for a reaction to occur.

✔ Equilibrium tells chemists how much product they'll get if they wait long enough.

These two concepts are completely separate. There's no connection between how productive a reaction can be and how long that reaction takes to proceed. In other words, chemists have good days and bad days, like everyone else. At least they have a little bit of theory to help them make sense of these things. In this chapter, you get an overview of this theory. Don't wait. Read on.

Measuring Rates

So you've got this beaker, and a reaction is going on inside it. Is the reaction a fast one or a slow one? How fast or how slow? How can you tell? These are questions about rates. You can measure a reaction rate by measuring how fast a reactant disappears or by measuring how fast a product appears. If the reaction occurs in solution, the molar concentration of reactant or product changes over time, so rates are often expressed in units of molarity per second (M/s).

For the following reaction,

$$A + B \rightarrow C$$

you can measure the reaction rate by measuring either the decrease in the concentration of Reactant A (or B) or the increase in the concentration of Product C over time:

$$\text{Rate} = -\frac{d[A]}{dt} = \frac{d[C]}{dt}$$

In these kinds of equations, d is math-speak for a change in the amount of something at any given moment.

If you plot the concentration of product against the reaction time, you might get a curve like the one in Figure 14-1. Reactions usually occur most quickly at the beginning of a reaction, when the concentration of products is the lowest and the concentration of reactants is highest. The precise rate at any given moment of the reaction is called the *instantaneous rate* and is equal to the slope of a line drawn tangent to the curve.

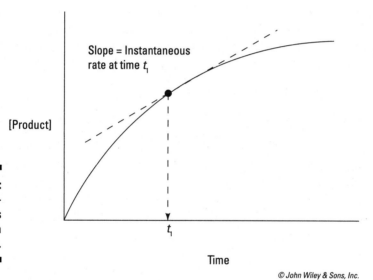

Figure 14-1:
The instantaneous rate of a reaction.

Slope = Instantaneous rate at time t_i

[Product]

t_i

Time

© John Wiley & Sons, Inc.

Clearly, the rate can change as the concentration of reactants and/or products changes. So you'd expect that any equation for describing the rate of a reaction like the one in Figure 14-1 would include some variable for the concentration of reactant. You'd be right.

Equations that relate the rate of a reaction to the concentration of some *species* (which can mean either reactant or product) in solution are called *rate laws*. The exact form that a rate law assumes depends on the reaction involved. Countless research studies have described the intricacies of rate in chemical reactions. Here, we focus on rate laws for simpler reactions. In general, rate laws take the following form:

$$\text{Rate} = k\left[\text{Reactant A}\right]^m\left[\text{Reactant B}\right]^n$$

This rate law describes a reaction whose rate depends on the concentration of two reactants, A and B. Other rate laws for other reactions may include factors for more or fewer reactants. In this equation, k is the *rate constant,* a number that must be experimentally measured for different reactions. The exponents *m* and *n* are called *reaction orders,* and they must also be measured for different reactions. A reaction order reflects the impact of a change in concentration in overall rate. If $m > n$, then a change in the concentration of A affects the rate more than does changing the concentration of B. The sum, $m + n$, is the overall reaction order.

REMEMBER

Some simple kinds of reaction rate laws crop up frequently, so they're worth your notice:

- ✔ **Zero-order reactions:** Rates for these reactions don't depend on the concentration of any species but simply proceed at a characteristic and constant rate.

$$Rate = k$$

- ✔ **First-order reactions:** Rates for these reactions typically depend on the concentration of a single species.

$$Rate = k[A]$$

- ✔ **Second-order reactions:** Rates for these reactions may depend on the concentration of two species, or they may have second-order dependence on the concentration of a single species (or some intermediate combination).

$$Rate = k[A][B] \quad \text{or} \quad Rate = k[A]^2$$

Measuring rates can help you figure out *mechanism,* the molecular details by which a reaction takes place. A *rate-determining step,* the step used to figure out the overall rate, is determined by identifying the *slowest* of the steps in the reaction mechanism. By measuring reaction rates under varying conditions, you can discover a lot about the chemical nature of the rate-determining step. Be forewarned, however: You can never use rates to prove a reaction mechanism; you can use rates only to disprove incorrect mechanisms.

EXAMPLE

Q. Consider the following two reactions and their associated rate laws:

| Reaction 1: | A+B → C | $Rate = k[A]^2$ |
| Reaction 2: | D+E → F+2G | $Rate = k[D][E]$ |

a. What is the overall reaction order for Reaction 1 and for Reaction 2?

b. For Reaction 1, how will the rate change if the concentration of A is doubled? How will the rate change if the concentration of B is doubled?

c. For Reaction 2, how will the rate change if the concentration of D is doubled? How will the rate change if the concentration of E is doubled?

d. What is the relationship between the rates of change in [A], [B], and [C]? What is the relationship between the rates of change in [D], [E], [F], and [G]?

A. Based on the given reactions and their associated rate laws, here are the answers:

a. Both reactions are second-order reactions, because adding the individual reaction orders yields 2 in each case.

b. If [A] is doubled, the rate law predicts that the rate will quadruple, because $2^2 = 4$. If [B] is doubled, the rate won't change, because [B] doesn't appear in the rate law.

c. If [D] is doubled or if [E] is doubled, the rate law predicts that the rate will also double, because $2^1 = 2$.

d. In Reaction 1, Reactants A and B are consumed to make Product C, so the rates of change in the concentrations of A and B will have the opposite sign as the rate of change in C (you could swap positive and negative values throughout the rate equation, and your answer would still be correct).

$$\text{Rate} = \frac{d[A]}{dt} = \frac{d[B]}{dt} = -\frac{d[C]}{dt}$$

In Reaction 2, the same guidelines apply for positive and negative signs, but there's an added wrinkle: 2 mol of

Product G are made for every 1 mol made of Product F and for every 1 mol consumed of Reactants D and E. So the rate of change in D, E, and F is one-half the rate of change in G, as indicated by the coefficient of 0.5.

$$\text{Rate} = \frac{d[D]}{dt} = \frac{d[E]}{dt} = -\frac{d[F]}{dt} = \frac{-0.5d[G]}{dt}$$

---

1. Methane combusts with oxygen to yield carbon dioxide and water vapor:

$$CH_4 + 2O_2 \rightarrow CO_2 + 2H_2O$$

If methane is consumed at 2.79 mol/s, what is the rate of production of carbon dioxide and water?

Solve It

2. You study the following reaction:

$$A + B \rightarrow C$$

You observe that tripling the concentration of A increases the rate by a factor of 9. You also observe that doubling the concentration of B doubles the rate. Write the rate law for this reaction. What is the overall reaction order?

Solve It

3. You study the following reaction:

$$D + E \rightarrow F + 2G$$

You vary the concentration of reactants D and E and observe the resulting rates:

	[D] in M	[E] in M	Rate in M/s
Trial 1:	2.7×10^{-2}	2.7×10^{-2}	4.8×10^{6}
Trial 2:	2.7×10^{-2}	5.4×10^{-2}	9.6×10^{6}
Trial 3:	5.4×10^{-2}	2.7×10^{-2}	9.6×10^{6}

Write the rate law for this reaction and calculate the rate constant, k. At what rate will the reaction occur in the presence of 1.3×10^{-2} M Reactant D and 0.92×10^{-2} M Reactant E?

Solve It

Focusing on Factors that Affect Rates

Chemists are finicky, tinkering types. They usually want to change reaction rates to suit their own needs. What can affect rates, and why? Temperature, concentration, and catalysts influence rate as follows:

✔ **Reaction rates tend to increase with temperature.** This trend results from the fact that reactants must collide with one another to have the chance to react. If reactants collide with the right orientation and with enough energy, the reaction can occur. So the greater the number of collisions and the greater the energy of those collisions, the more actual reacting takes place. An increase in temperature corresponds to an increase in the average kinetic energy of the particles in a reacting mixture — the particles move faster, colliding more frequently and with greater energy. (See Chapter 10 for info on kinetic energy.)

✔ **Increasing concentration tends to increase the reaction rate.** The reason for this trend also has to do with collisions. A higher concentration means that more reactant particles are closer together, so they undergo more collisions and have a greater chance of reacting. Increasing the concentration of reactants may mean dissolving more of those reactants in solution.

Some reactants aren't completely dissolved but come in larger, undissolved particles. In these cases, smaller particles lead to faster reactions. Smaller particles expose more surface area, making a greater portion of the particle available for reaction.

✔ **Catalysts increase reaction rates.** Catalysts don't themselves become chemically changed, and they don't alter the amount of product a reaction can eventually produce (the *yield*). An example from early childhood fits here. When you're learning to ride a bike, you might get a push from your parents to help you get going. However, after that push, the pedaling is entirely up to you. Your top speed and end destination are still entirely regulated by your ability to pedal and steer the bike, but that push (the catalyst) helped you get up to speed more quickly.

Catalysts can operate in many different ways, but all those ways have to do with decreasing *activation energy,* the energetic hill reactants must climb to reach a *transition state,* the highest-energy state along a reaction pathway. Lower activation energies mean faster reactions. Figure 14-2 shows a *reaction progress diagram,* the energetic pathway that reactants must traverse to become products.

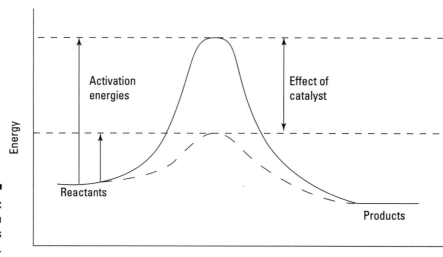

Figure 14-2:
A reaction
progress
diagram.

© *John Wiley & Sons, Inc.*

Q. Consider the following reaction:

$$H_2(g) + Cl_2(g) \rightarrow 2HCl(g)$$

If 1 mol of H_2 reacts with 1 mol of Cl_2 to form 2 mol of HCl, does the reaction occur more rapidly in a 5 L vessel or a 10 L vessel? Does it occur more rapidly at 273 K or 293 K? Why?

A. The reaction occurs more rapidly in the 5 L vessel because the concentration of reactant molecules is higher when they occupy the smaller volume. At higher concentrations, more collisions occur between reactant molecules. At higher temperatures, particles move with greater energy, which also produces more collisions and collisions of greater force. So the reaction occurs more rapidly at 293 K.

4. Here is a simplified reaction equation for the combustion of gunpowder:

$$10KNO_3 + 3S + 8C \rightarrow 2K_2CO_3 + 3K_2SO_4 + 6CO_2 + 5N_2$$

In the 1300s, powdersmiths began to process raw gunpowder by using a method called *corning*. Corning involves adding liquid to gunpowder to make a paste, pressing the paste into solid cakes, and forcing the cakes through a sieve to produce grains of defined size. These grains are both larger and more consistently sized than the original grains of powder. Explain the advantages of corned gunpowder from the perspective of chemical kinetics.

Solve It

5. Consider the following reaction:

$$A \rightarrow B$$

To progress from reactant to product, Reactant A must pass through a high-energy transition state, A*. Imagine that reaction conditions are changed in such a way that the energy of A is increased, the energy of B is decreased, and the energy of A* remains unchanged. Does the change in conditions result in a faster reaction? Why or why not?

Solve It

Measuring Equilibrium

Figure 14-3 shows a reaction progress diagram like the one in Figure 14-2 but highlights the energy difference between reactants and products. This difference is completely independent of activation energy, which we discuss in the preceding section. Although activation energy controls the rate of a reaction, the difference in energy between reactants and products determines the extent of a reaction — how much reactant will have converted to product when the reaction is complete.

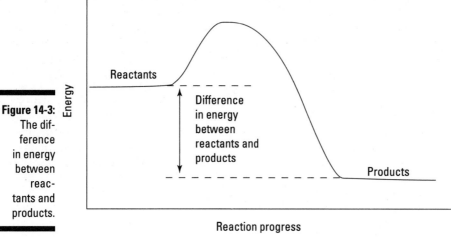

© John Wiley & Sons, Inc.

A reaction that has produced as much product as it ever will is said to be at *equilibrium*. Equilibrium doesn't mean that no more chemistry is occurring; rather, equilibrium means that the rate at which reactant converts to product equals the rate at which product converts back into reactant. (Yes, reactions go both forward and backward at all times.) So the overall concentrations of reactants and products no longer change.

Two key ideas emerge:

- ✔ The equilibrium position of a reaction (the extent to which the reaction proceeds) can be characterized by the concentrations of reactants and products, because these concentrations no longer change.

- ✔ The equilibrium position of a reaction is intimately connected to the difference in energy between reactants and products.

These two ideas are expressed by two important equations, which we cover next.

The equilibrium constant

A parameter called K_{eq}, the *equilibrium constant*, describes the equilibrium position of a reaction in terms of the concentrations of reactants and products. For the reaction

$$aA + bB \leftrightarrow cC + dD$$

the equilibrium constant is calculated as follows:

$$K_{eq} = \frac{[C]^c[D]^d}{[A]^a[B]^b}$$

Figure 14-3: The difference in energy between reactants and products.

In general, concentrations of the products are divided by the concentrations of the reactants. In the case of gas-phase reactions, partial pressures are used instead of molar concentrations. Multiple product or reactant concentrations are multiplied. Each concentration is raised to an exponent equal to its stoichiometric coefficient in the balanced reaction equation. (See Chapters 8 and 9 for details on balanced equations and stoichiometry.)

Very favorable reactions (ones that progress on their own) produce a lot of product, so they have K_{eq} values much larger than 1. Very unfavorable reactions (ones that require an input of energy) convert very little reactant into product, so they have K_{eq} values between 0 and 1. In a reaction with $K_{eq} = 1$, the amount of product equals the amount of reactant at equilibrium.

Note that you can calculate K_{eq} only by using concentrations measured at equilibrium. Concentrations measured before a reaction reaches equilibrium are used to calculate a *reaction quotient, Q*, which allows you determine the general direction of the reaction at the moment:

$$Q = \frac{[C]^c[D]^d}{[A]^a[B]^b}$$

If $Q < K_{eq}$, the reaction will progress to the right, making more product. If $Q > K_{eq}$, the reaction will shift to the left, converting product into reactant. If $Q = K_{eq}$, the reaction is at equilibrium.

Free energy

The second key equation having to do with equilibrium relates K_{eq} to the difference in energy between reactants and products. The particular form of energy important in this relationship is *free energy, G*. The difference in free energy between product and reactant states is $\Delta G = G_{product} - G_{reactant}$. The relationship between free energy and equilibrium is

$$\Delta G = -RT \ln K_{eq} \quad \text{or} \quad K_{eq} = e^{\left(\frac{-\Delta G}{RT}\right)}$$

In these equations, R is the gas constant (0.08206 L·atm/(mol·K) or 8.314 L·kPa/(mol·K), depending on your units; see Chapter 11 for more information), T is temperature, and *ln* refers to the natural logarithm (log base *e*). The equation is typically true for reactions that occur with no change in temperature or pressure.

Favorable reactions possess negative values for ΔG, and unfavorable reactions possess positive values for ΔG. Energy must be added to drive an unfavorable reaction forward. If the ΔG for a set of reaction conditions is 0, the reaction is at equilibrium.

The free energy change for a reaction, ΔG, arises from the interplay of two other parameters, the *enthalpy* change, ΔH, and the *entropy* change, ΔS. A detailed discussion of enthalpy and entropy is well beyond the scope of this book and would in truth fill several books of their own. As a *rough* approximation, you can think of enthalpy as energy in the system and of entropy as a measure of disorder in the system (when describing enthalpy in tandem with entropy, you can think of enthalpy as a measure of *order* in a system).

TIP

Thankfully, we can somewhat explain enthalpy and entropy from a commonsense standpoint. The universe tends toward disorder and a loss of organized energy at every turn. Your everyday life is no exception. As a very simple example, imagine your bedroom. If you change your clothes, you have two choices: Spend less energy and simply leave the clothes where you took them off, or spend more energy and put them in the laundry basket. Sadly, without the input of organized energy, your room will not get clean on its own. Reactions function in much the same way: They naturally tend toward a state of disorder at the expense of organized energy. That's why a negative change in enthalpy (organized energy) and a positive change in entropy (disorder and the loss of organized energy) are always favorable directions for a reaction. Overall, then

$$\Delta G = \Delta H - T\Delta S$$

where T is temperature.

EXAMPLE

Q. Consider the following reaction:

$$A + 2B \leftrightarrow 2C$$

a. Write the expression that relates the equilibrium constant for this reaction to the concentrations of the reactants and product.

b. If the $K_{eq} = 1.37 \times 10^3$, what is the free energy change for the reaction at 298 K?

c. At equilibrium, are reactants or product favored?

Note: $R = 8.314 \dfrac{\text{L} \cdot \text{kPa}}{\text{mol} \cdot \text{K}} = 8.314 \dfrac{\text{J}}{\text{mol} \cdot \text{K}}$

A. Here are answers for the questions about the reaction $A + 2B \leftrightarrow 2C$:

a. Remember that the expression for the reaction is always products over reactants. Remember also to use the coefficients as exponents for each respective concentration.

$$K_{eq} = \frac{[C]^2}{([A][B]^2)}$$

b. **−17.9 kJ/mol.** Enter the given values of R, T, and K_{eq} into the ΔG equation:

$$\Delta G = -RT \ln K_{eq}$$

$$= -\left(8.314 \ \frac{\text{J}}{\text{mol} \cdot \text{K}}\right)(298 \text{ K})\left(\ln 1.37 \times 10^3\right)$$

$$= -1.79 \times 10^4 \ \frac{\text{J}}{\text{mol}}$$

$$= -17.9 \ \frac{\text{kJ}}{\text{mol}}$$

c. $K_{eq} > 1$ indicates the product is favored, and "at equilibrium" indicates that $\Delta G = 0$. No more changes will be taking place, so **the product is favored.**

6. Consider the following reaction:

$$N_2(g) + O_2(g) \leftrightarrow 2NO(g)$$

At equilibrium, you measure the following partial pressures:

$$P(N_2) = 4.76 \times 10^{-2} \text{ atm}$$
$$P(O_2) = 9.82 \times 10^{-3} \text{ atm}$$
$$P(NO) = 2.63 \times 10^{-7} \text{ atm}$$

a. What is the K_{eq} for this reaction?

b. If you measured $P(O_2) = 3.74 \times 10^{-2}$ atm and if other partial pressures remained unchanged, what is the reaction quotient for the reaction? In which direction would the reaction proceed?

Solve It

7. You perform the following reaction:

$$2A + 2B \leftrightarrow 3C + D$$

After waiting three hours, you measure the following concentrations:

$$[A] = 273 \text{ mM}$$
$$[B] = 34.7 \text{ mM}$$
$$[C] = 0.443 \text{ M}$$
$$[D] = 78.9 \text{ mM}$$

a. What is the reaction quotient for the system?

b. If the K_{eq} for this reaction is 1.85×10^2, has the reaction completed?

Solve It

Answers to Questions on Rates and Equilibrium

You may have raced through these problems, or you may have moved at the speed of a dead snail in winter. But rate is separate from equilibrium — whatever your pace, you've made it this far. Now shift in opposition to any perturbing problems; check your work.

1 **CO_2: 2.79 mol/s; H_2O: 5.58 mol/s.** The reaction equation makes clear that each mole of methane reactant corresponds to 1 mol of carbon dioxide product and 2 mol of water product. So

$$\frac{d[CH_4]}{dt} = -\frac{d[CO_2]}{dt} = -0.5\frac{d[H_2O]}{dt}$$

This means that the disappearance of 2.79 mol/s of methane corresponds to the appearance of 2.79 mol/s of CO_2 and the appearance of $2 \times 2.79 = 5.58$ mol/s of H_2O. The concentration of methane changes at half the rate of water, so the concentration of water changes at twice the rate of methane.

2 **Third-order.** The observations are consistent with the rate law, Rate $= k[A]^2[B]$. Tripling [A] increases the rate ninefold, and $3^2 = 9$. Doubling [B] doubles the rate, and $2^1 = 2$. The overall reaction order (3) is the sum of the individual reaction orders (the exponents on the individual reactant concentrations): $2 + 1 = 3$.

3 Doubling the concentration of either reactant (D or E) doubles the rate. So the data is consistent with the rate law, Rate $= k[D][E]$. Solve for k by substituting known values of rates [D] and [E] from any of the trial reactions:

$$k = \frac{\text{Rate}}{[D][E]} = \frac{4.8 \times 10^6 \text{ M/s}}{(2.7 \times 10^{-2} \text{ M})(2.7 \times 10^{-2} \text{ M})} = 6.6 \times 10^9 \text{ M/s}$$

Use this calculated value of k to determine the rate in the presence of 1.3×10^{-2} M Reactant D and $0.92 \cdot 10^{-2}$ M Reactant E:

$$\text{Rate} = (6.6 \times 10^9 \text{ M/s})(1.3 \times 10^{-2} \text{ M})(0.92 \times 10^{-2} \text{ M}) = 7.9 \times 10^5 \text{ M/s}$$

4 The corning process converts gunpowder from a finely divided powder of undefined particle size into larger particles of defined size. Because the corned particles are larger, combustion occurs more slowly. Corned gunpowder is less likely to explode accidentally than is a fine dust of gunpowder. Because the corned particles are all the same size, combustion occurs at a predictable rate, which is convenient for the user.

5 The change in conditions does result in a faster rate. Rate is limited by activation energy, the difference in energy between the reactant (A) and the transition state (A*). If the energy of A is raised and the energy of A* remains constant, then the difference in energy (the activation energy) becomes smaller, and lower activation energies result in faster reactions. The decrease in the energy of Product B has no bearing on the rate because it has no effect on the activation energy for the A → B reaction — although it would affect the rate of the reverse reaction, B → A.

6 Here are the answers regarding equilibrium in the reaction $N_2\,(g) + O_2\,(g) \leftrightarrow 2NO\,(g)$:

a. 1.48×10^{-10}. You can find the K_{eq} for the reaction by setting up a simple equilibrium expression. Remember to put product pressures over reactant pressures and to apply the coefficients as exponents to each one. In this example, only NO has a coefficient of more than 1, so you square the given pressure of NO.

$$K_{eq} = \frac{[2.63 \times 10^{-7}\ \text{atm}]^2}{([4.76 \times 10^{-2}\ \text{atm}]\,[9.82 \times 10^{-3}\ \text{atm}])} = 1.48 \times 10^{-10}$$

b. To solve for Q (the reaction quotient), your formula is almost exactly the same as the equilibrium expression in part (a). The only change is that the problem gives you a new pressure for O_2, so instead of using the value given at equilibrium, you use the new value.

$$Q = \frac{[2.63 \times 10^{-7}\ \text{atm}]^2}{([4.76 \times 10^{-2}\ \text{atm}]\,[3.74 \times 10^{-2}\ \text{atm}])} = 3.89 \times 10^{-11}$$

Because $Q < K_{eq}$, the reaction proceeds to the right, converting reactant into product.

7 To solve this problem, you first need to determine Q and then compare it to the K_{eq} given to you in the problem. To determine Q, set up your equilibrium expression with products over reactants and coefficients applied as exponents. After you get your Q, compare it to the K_{eq}. If K_{eq} is larger, the reaction will continue.

a. 76.4

$$Q = \frac{([78.9\ \text{mM}][443\ \text{mM}]^3)}{([273\ \text{mM}]^2[34.7\ \text{mM}]^2)} = 76.4$$

b. No. If $K_{eq} = 1.85 \times 10^2$, then $Q < K_{eq}$, so the reaction will continue to convert reactant to product.

Chapter 15

Warming Up to Thermochemistry

. . . .

In This Chapter

▶ Getting a brief overview of thermodynamics

▶ Using heat capacity and calorimetry to measure heat flow

▶ Keeping track of the heats involved in chemical and physical changes

▶ Adding heats together with Hess's law

. . . .

*E*nergy shifts between many forms. It may be tricky to detect, but energy is always conserved. Sometimes energy reveals itself as heat. *Thermodynamics* explores how energy
moves from one form to another. *Thermochemistry* investigates changes in heat that accompany chemical reactions. In this chapter, we delve into the particulars of thermodynamics
and thermochemistry.

Understanding the Basics of Thermodynamics

To understand how thermochemistry is done, you need to first understand how the particular form of energy called *heat* fits into the overall dance of energy and matter.

To study energy, it helps to divide the universe into two parts, the *system* and the *surroundings*. For chemists, the system may consist of the contents of a beaker or tube. This is an
example of a *closed system,* one that allows exchange of energy with the surroundings but
doesn't allow exchange of matter. Closed systems are common in chemistry. Though energy
may move between system and surroundings, the total energy of the universe is constant.

Energy itself is divided into potential energy and kinetic energy:

▶ **Potential energy (PE):** Potential energy is energy due to position. *Chemical energy* is a
kind of potential energy arising from the positions of particles within systems and from
the energy stored within bonds.

▶ **Kinetic energy (KE):** Kinetic energy is the energy of motion. *Thermal energy* is a kind of
kinetic energy, arising from the movement of particles within systems.

The total *internal energy* of a system *(E)* is the sum of its potential and kinetic energies. When a system moves between two states (as it does in a chemical reaction), the internal energy may change as the system exchanges energy with the surroundings. The difference in energy (ΔE) between the initial and final states derives from heat *(q)* added to or lost from the system and from work *(w)* done by the system or on the system.

We can summarize these energy explanations with the help of a couple of handy formulas:

$$E_{total} = KE + PE$$

$$\Delta E = E_{final} - E_{initial} = q + w$$

What kind of work can atoms and molecules do in a chemical reaction? One kind that's easy to understand is *pressure-volume work.* Consider the following reaction:

$$CaCO_3(s) \rightarrow CaO(s) + CO_2(g)$$

Solid calcium carbonate decomposes into solid calcium oxide and carbon dioxide gas. At constant pressure *(P)*, this reaction proceeds with a change in volume *(V)*. The added volume comes from the production of carbon dioxide gas. As gas is made, it expands, pushing against the surroundings. The carbon dioxide gas molecules do work as they push into a greater volume:

$$w = -P\Delta V$$

The negative sign in this equation means that the system loses internal energy because of the work it does on the surroundings. If the surroundings did work on the system, thereby decreasing the system's volume, then the system would gain internal energy.

So pressure-volume work can partly account for changes in internal energy during a reaction. When pressure-volume work is the only kind of work involved, any remaining changes come from heat flow. *Enthalpy (H)* corresponds to the heat content of a closed system at constant pressure. An *enthalpy change (ΔH)* in such a system corresponds to heat flow. The enthalpy change equals the change in internal energy minus the energy used to perform pressure-volume work:

$$\Delta H = \Delta E - (-P\Delta V) = \Delta E + P\Delta V$$

Like *E, P,* and *V, H* is a *state function,* meaning that its value has only to do with the state of the system and nothing to do with how the system got to that state. Heat *(q)* is *not* a state function; it's simply a form of energy that flows from warmer objects to cooler objects.

Now breathe. The practical consequences of all this theory are the following:

- Chemical reactions usually involve the flow of energy in the form of heat, *q.*
- Chemists monitor changes in heat by measuring changes in temperature.
- At constant pressure, the change in heat content equals the change in enthalpy, *ΔH.*
- Knowing ΔH values helps you explain and predict chemical behavior.

Q. Gas is heated within a sealed cylinder. The heat causes the gas to expand, pushing a movable piston outward to increase the volume of the cylinder to 4.63 L. The initial and final pressures of the system are both 1.15 atm. The gas does 304 J of work on the piston. What was the initial volume of the cylinder? (*Note:* 101.3 J = 1 L·atm)

A. **2.02 L.** You're given an amount of work, a constant pressure, and the knowledge that the volume of the system changes. So the equation to use here is $w = -P\Delta V = -P\left(V_{final} - V_{initial}\right)$. Because the system does work on the piston (and not the other way around), the sign of w is negative. Substituting your known values into the equation gives you

$$-304 \text{ J} = -\left(1.15 \text{ atm}\right)\left(4.63 \text{ L} - V_{initial}\right)$$

The units don't match, so convert joules to liter-atmospheres:

$$(304 \text{ J})\left(\frac{1 \text{ L·atm}}{101.3 \text{ J}}\right) = 3.00 \text{ L·atm}$$

Now you can solve for the initial volume:

$$-3.00 \text{ L·atm} = (-1.15 \text{ atm})\left(4.63 \text{ L} - V_{initial}\right)$$

$$\frac{-3.00 \text{ L·atm}}{-1.15 \text{ atm}} = \frac{(-1.15 \text{ atm})\left(4.63 \text{ L} - V_{initial}\right)}{-1.15 \text{ atm}}$$

$$2.61 \text{ L} = 4.63 \text{ L} - V_{initial}$$

$$V_{initial} = 2.02 \text{ L}$$

---

1. A fuel combusts at 3.00 atm constant pressure. The reaction releases 75.0 kJ of heat and causes the system to expand from 7.50 L to 20.0 L. What is the change in internal energy? (*Note:* 101.3 J = 1 L·atm)

Solve It

Working with Specific Heat Capacity and Calorimetry

Heat is a form of energy that flows from warmer objects to cooler objects. But how much heat can an object hold? If objects have the same heat content, does that mean they're the same temperature? You can measure different temperatures, but how do these temperatures relate to heat flow? These kinds of questions revolve around the concept of *heat capacity,* the amount of heat required to raise the temperature of a system by 1°C, or 1 K.

You'll encounter heat capacity in different forms, each of which is useful in different scenarios. Any system has a heat capacity. But how can you best compare heat capacities between chemical systems? You use *molar heat capacity* or *specific heat capacity* (or just *specific heat*). Molar heat capacity is simply the heat capacity of 1 mol of a substance. Specific heat capacity is simply the heat capacity of 1 g of a substance. How do you know whether you're dealing with heat capacity, molar heat capacity, or specific heat capacity? Look at the units.

- ✔ Heat capacity: $\dfrac{\text{Energy}}{\text{K}}$

- ✔ Molar heat capacity: $\dfrac{\text{Energy}}{\text{mol} \cdot \text{K}}$

- ✔ Specific heat capacity: $\dfrac{\text{Energy}}{\text{g} \cdot \text{K}}$

The number representing specific heat describes the amount of energy required to raise 1 g of a substance by 1°C, or 1 K. If you have a larger amount of a substance, raising the temperature will take longer. To give you a practical and familiar example, imagine two pots of water, one big and one small, coming to a boil on a stove. The big one will take longer to warm up compared to the small one.

Fine, but what are the units of energy? Well, that depends. The SI unit of energy is the *joule* (J), but the *calorie* (cal) and *liter-atmosphere* (L·atm) are also used. Here's how the joule, the calorie, and the liter-atmosphere are related:

- ✔ 1 J = 0.2390 cal
- ✔ 101.3 J = 1 L·atm

Note that a food "calorie" actually refers to a *Calorie,* which is a *kilocalorie* — a calorie of cheesecake is 1,000 times larger than you think it is.

Calorimetry is a family of techniques that puts all this thermochemical theory to use. When chemists do calorimetry, they initiate a reaction within a defined system and then measure any temperature change that occurs as the reaction progresses. There are a few variations on this theme:

- ✔ **Constant-pressure calorimetry:** Constant-pressure calorimetry directly measures an enthalpy change (ΔH) for a reaction because it monitors heat flow at constant pressure: $\Delta H = q_p$.

  Typically, heat flow is observed through changes in the temperature of a reaction solution. If a reaction warms a solution, then that reaction must have released heat into the solution. In other words, the change in heat content of the reaction (q_{reaction}) has the same magnitude as the change in heat content for the solution (q_{solution}) but has the opposite sign: $q_{\text{solution}} = -q_{\text{reaction}}$.

  Measuring q_{solution} allows you to calculate q_{reaction}, but how can you measure q_{solution}? You do so by measuring the difference in temperature (ΔT) before and after the reaction:

  $$q_{\text{solution}} = (\text{mass of solution})(\text{specific heat of solution})(\Delta T)$$

  In other words, $q = mC_p\Delta T$. Here, m is the mass of the solution and C_p is the specific heat capacity of the solution at constant pressure. ΔT is equal to $T_{\text{final}} - T_{\text{initial}}$.

WARNING!

When you use this equation, be sure that all your units match. For example, if your C_p has units of J/(g·K), don't expect to calculate heat flow in kilocalories. A common source of error in solving specific heat problems is the need to use the correct temperature units; be sure to pay attention to whether your temperature is in kelvins or degrees Celsius.

✔ **Constant-volume calorimetry:** Constant-volume calorimetry directly measures a change in internal energy (ΔE, not ΔH) for a reaction because it monitors heat flow at constant volume. Often, ΔE and ΔH are very similar values.

TIP

A common variety of constant-volume calorimetry is *bomb calorimetry*, a technique in which a reaction (often, a combustion reaction) is triggered within a sealed vessel called a *bomb.* The vessel is immersed in a water bath of known volume. The temperature of the water is measured before and after the reaction. Because the heat capacity of the water and the calorimeter are known, you can calculate heat flow from the change in temperature.

EXAMPLE

Q. Paraffin wax is sometimes incorporated in sheetrock to act as an insulator. During the day, the wax absorbs heat and melts. During the cool nights, the wax releases heat and solidifies. At sunrise, a small hunk of solid paraffin within a wall has a temperature of 298 K. The rising sun warms the wax, which has a melting temperature of 354 K. If the hunk of wax has a mass of 0.257 g and a specific heat capacity of 2.50 J/(g·K), how much heat must the wax absorb to bring it to its melting point?

A. **36.0 J.** You're given two temperatures, a mass, and a specific heat capacity. You're asked to find an amount of heat energy. You have all you need to proceed with $q = mC_p\Delta T$:

$$q = (0.257 \text{ g}) \left(2.50\frac{E}{g \cdot K}\right) (354 \text{ K} - 298 \text{ K}) = 36.0 \text{ J}$$

2. A 375 g plug of lead is heated and placed in an insulated container filled with 0.500 L of water. Prior to the immersion of the lead, the water is at 293 K. After a time, the lead and the water assume the same temperature, 297 K. The specific heat capacity of lead is 0.127 J/(g·K), and the specific heat capacity of water is 4.18 J/(g·K). How hot was the lead before it entered the water? (**Hint:** You'll need to use the density of water.)

Solve It

3. At some point, all laboratory chemists learn the same hard lesson: Hot glass looks just like cold glass. Heath discovered this when he picked up a hot beaker someone left on his bench. At the moment Heath grasped the 413 K glass beaker, 567 J of heat flowed out of the beaker and into his hand. The glass of the beaker has a heat capacity of 0.84 J/(g·K). If the beaker was 410 K the instant Heath dropped it, then what is the collective mass of the shards of beaker now littering the lab floor?

Solve It

4. 25.4 g of sodium hydroxide (NaOH) is dissolved in water within an insulated calorimeter. The heat capacity of the resulting solution is 4.18×10^3 J/K. The temperature of the water prior to the addition of NaOH was 296 K. If NaOH releases 44.2 kJ/mol as it dissolves, what is the final temperature of the solution?

Solve It

Absorbing and Releasing Heat: Endothermic and Exothermic Reactions

You can monitor heat flow by measuring changes in temperature, but what does any of this have to do with chemistry? Chemical reactions transform both matter and energy. Though reaction equations usually list only the matter components of a reaction, you can also consider heat energy as a reactant or product. When chemists are interested in heat flow during a reaction (and when the reaction is run at constant pressure), they may list an enthalpy change (ΔH) to the right of the reaction equation. As we explain in the preceding section, at constant pressure, heat flow equals ΔH:

$$q_p = \Delta H = H_{final} - H_{initial}$$

If the ΔH listed for a reaction is negative, then that reaction releases heat as it proceeds — the reaction is *exothermic* (*exo-* = out). If the ΔH listed for the reaction is positive, then that reaction absorbs heat as it proceeds — the reaction is *endothermic* (*endo-* = in). In other words, exothermic reactions release heat as a product, and endothermic reactions consume heat as a reactant.

The sign of the ΔH tells you the direction of heat flow, but what about the magnitude? The coefficients of a chemical reaction represent molar equivalents (see Chapter 8 for details), so the value listed for the ΔH refers to the enthalpy change for one molar equivalent of the reaction. Here's an example:

$$CH_4(g) + 2\,O_2(g) \rightarrow CO_2(g) + 2H_2O(g) \quad \Delta H = -802 \text{ kJ}$$

This reaction equation describes the combustion of methane, a reaction you might expect to release heat. The enthalpy change listed for the reaction confirms this expectation: For each mole of methane that combusts, 802 kJ of heat is released. The reaction is highly exothermic. Based on the stoichiometry of the equation, you can also say that 802 kJ of heat is released for every 2 mol of water produced. (Flip to Chapter 9 for the scoop on stoichiometry.)

So reaction enthalpy changes (or reaction "heats") are a useful way to measure or predict chemical change. But they're just as useful in dealing with physical changes, like freezing and melting, evaporating and condensing, and others. For example, water (like most substances) absorbs heat as it melts (or *fuses*) and as it evaporates. Here are the molar enthalpies for such changes:

REMEMBER

☞ **Molar enthalpy of fusion:** $\Delta H_{fus} = 6.01$ kJ

☞ **Molar enthalpy of vaporization:** $\Delta H_{vap} = 40.68$ kJ

The same sorts of rules apply to enthalpy changes listed for chemical changes and physical changes. Here's a summary of the rules that apply to both:

☞ **The heat absorbed or released by a process is proportional to the moles of substance that undergo that process.** For example, 2 mol of combusting methane release twice as much heat as 1 mol of combusting methane.

☞ **Running a process in reverse produces heat flow of the same magnitude but of opposite sign as running the forward process.** For example, freezing 1 mol of water releases the same amount of heat that is absorbed when 1 mol of water melts.

EXAMPLE

Q. Here is a balanced chemical equation for the oxidation of hydrogen gas to form liquid water, along with the corresponding enthalpy change:

$$2H_2(g) + O_2(g) \rightarrow 2H_2O(l) \quad \Delta H = -572 \text{ kJ}$$

How much electrical energy must be expended to perform electrolysis of 3.76 mol of liquid water, converting that water into hydrogen gas and oxygen gas?

A. **1.08×10^3 kJ.** First, recognize that the given enthalpy change is for the reverse of the electrolysis reaction, so you must reverse its sign from –572 to 572. Second, recall that heats of reaction are proportional to the amount of substance reacting (2 mol of H_2O in this case), so the calculation is

$$(3.76 \text{ mol } H_2O) \left(\frac{572 \text{ kJ}}{2 \text{ mol } H_2O} \right) = 1.08 \times 10^3 \text{ kJ}$$

---

5. Carbon dioxide gas can be decomposed into oxygen gas and carbon monoxide:

$$2 CO_2(g) \rightarrow O_2(g) + 2CO(g) \quad \Delta H = 486 \text{ kJ}$$

How much heat is released or absorbed when 9.67 g of carbon monoxide combines with oxygen to form carbon dioxide?

Solve It

6. How much heat must be added to convert 4.77 mol of 268 K ice into steam? The specific heat capacity of ice is 38.1 J/(mol·K). The specific heat capacity of water is 75.3 J/(mol·K). The molar enthalpies of fusion and vaporization are 6.01 kJ and 40.68 kJ, respectively.

Solve It

Summing Heats with Hess's Law

For the chemist, *Hess's law* is a valuable tool for dissecting heat flow in complicated, multistep reactions. For the confused or disgruntled chemistry student, Hess's law is a breath of fresh air. In essence, the law confirms that heat behaves the way we'd like it to behave: predictably.

Imagine that the product of one reaction serves as the reactant for another reaction. Now imagine that the product of the second reaction serves as the reactant for a third reaction. What you have is a set of coupled reactions, connected in series like the cars of a train:

$$A \rightarrow B \quad \text{and} \quad B \rightarrow C \quad \text{and} \quad C \rightarrow D$$

Therefore,

$$A \rightarrow B \rightarrow C \rightarrow D$$

You can think of these three reactions adding up to one big reaction $A \rightarrow D$. What is the overall enthalpy change associated with this reaction $(\Delta H_{A \rightarrow D})$? Here's the good news:

$$\Delta H_{A \rightarrow D} = \Delta H_{A \rightarrow B} + \Delta H_{B \rightarrow C} + \Delta H_{C \rightarrow D}$$

Enthalpy changes are additive. But the good news gets even better. Imagine that you're trying to figure out the total enthalpy change for the following multistep reaction:

$$X \rightarrow Y \rightarrow Z$$

Here's a wrinkle: For technical reasons, you can't measure this enthalpy change $(\Delta H_{X \rightarrow Z})$ directly but must calculate it from tabulated values for $(\Delta H_{X \rightarrow Y})$ and $(\Delta H_{Y \rightarrow Z})$. No problem, right? You simply look up the tabulated values and add them. But here's another wrinkle: when you look up the tabulated values, you find the following:

$$\Delta H_{X \rightarrow Y} = -37.5 \frac{kJ}{mol}$$

$$\Delta H_{Z \rightarrow Y} = -10.2 \frac{kJ}{mol}$$

Gasp! You need $\Delta H_{Y \rightarrow Z}$, but you're provided only $\Delta H_{Z \rightarrow Y}$! Relax. The enthalpy change for a reaction has the same magnitude and opposite sign as the reverse reaction. So if $\Delta H_{Z \rightarrow Y} = -10.2$ kJ/mol, then $\Delta H_{Y \rightarrow Z} = 10.2$ kJ/mol. It really is that simple:

$$\Delta H_{X \rightarrow Z} = \Delta H_{X \rightarrow Y} + \left(-\Delta H_{Z \rightarrow Y} \right)$$

$$= -37.5 \frac{kJ}{mol} + 10.2 \frac{kJ}{mol} = -27.3 \frac{kJ}{mol}$$

Thanks be to Hess.

Q. Calculate the reaction enthalpy for the following reaction:

$$PCl_5(g) \rightarrow PCl_3(g) + Cl_2(g)$$

Use the following data:

Reaction 1:	$4PCl_3(g) \rightarrow P_4(s) + 6Cl_2(g)$	$\Delta H = 821 \text{ kJ}$
Reaction 2:	$P_4(s) + 10Cl_2(g) \rightarrow 4PCl_5(g)$	$\Delta H = -1,156 \text{ kJ}$

A. **83.8 kJ.** Reaction enthalpies are given for two reactions. Your task is to manipulate and add Reactions 1 and 2 so the sum is equivalent to the target reaction. First, reverse Reactions 1 and 2 to obtain Reactions 1′ and 2′, and add the two reactions. Identical species that appear on opposite sides of the equations cancel out (as occurs with species P_4 and Cl_2):

Reaction 1′:	$P_4(s) + 6Cl_2(g) \rightarrow 4PCl_3(g)$	$\Delta H = -821 \text{ kJ}$
Reaction 2′:	$4PCl_5(g) \rightarrow P_4(s) + 10Cl_2(g)$	$\Delta H = 1,156 \text{ kJ}$
Sum:	$4PCl_5(g) \rightarrow 4PCl_3(g) + 4Cl_2(g)$	$\Delta H = 335 \text{ kJ}$

Finally, divide the sum by 4 to yield the target reaction equation:

$$PCl_5(g) \rightarrow PCl_3(g) + Cl_2(g) \qquad \Delta H = 83.8 \text{ kJ}$$

---

7. Calculate the reaction enthalpy for the following reaction:

$$N_2O_4(g) \rightarrow N_2(g) + 2O_2(g)$$

Use the following data:

Reaction 1:	$N_2(g) + 2O_2(g) \rightarrow 2NO_2(g)$	$\Delta H = 1,032 \text{ kJ}$
Reaction 2:	$2NO_2(g) \rightarrow N_2O_4(g)$	$\Delta H = -886 \text{ kJ}$

Solve It

Answers to Questions on Thermochemistry

Check your answers to the practice problems to see whether you've truly felt the heat.

1 **78.8 kJ decrease.** You're given a constant pressure, an amount of heat, a change in volume, and the knowledge that there's been a change in internal energy. Changes in heat content at constant pressure are equivalent to a change in enthalpy, so the equation to use here is

$$\Delta H = \Delta E + P\Delta V$$

To do the math properly, you must make sure that all your units match, so convert the given heat energy from kJ to L·atm:

$$75.0 \text{ kJ} = 7.50 \times 10^4 \text{ J} = \left(7.50 \times 10^4 \text{ J}\right) \left(\frac{1 \text{ L·atm}}{101.3 \text{ J}}\right) = 7.40 \times 10^2 \text{ L·atm}$$

Because heat is released from the system, the change in enthalpy (ΔH) is negative. Substitute your known values into the equation:

$$-7.40 \times 10^2 \text{ L·atm} = \Delta E + (3.00 \text{ atm})(20.0 \text{ L} - 7.50 \text{ L})$$
$$-7.40 \times 10^2 \text{ L·atm} = \Delta E + (3.00 \text{ atm})(12.5 \text{ L})$$
$$-7.40 \times 10^2 \text{ L·atm} = \Delta E + (37.5 \text{ L·atm})$$
$$-778 \text{ L·atm} = \Delta E$$

Solving for ΔE gives you –778 L·atm, which is equivalent to –78.8 kJ. Because the sign is negative, the internal energy of the system decreases by 78.8 kJ.

2 **473 K.** The key to setting up this problem is to realize that whatever heat flows out of the lead flows into the water, so $q_{lead} = q_{water}$. Calculate each quantity of heat by using $q = mC_p\Delta T$. Recall that $\Delta T = T_{final} - T_{initial}$. The unknown in the problem is the initial temperature of lead:

$$q_{lead} = (375 \text{ g}) \left(0.127 \, \frac{\text{J}}{\text{g·K}}\right) \left(297 \text{ K} - T_{initial}\right)$$

To calculate q_{water}, you must first calculate the mass of 0.500 L water by using the density of water:

$$(0.500 \text{ L}) \left(1.00 \, \frac{\text{kg}}{\text{L}}\right) = 0.500 \text{ kg} = 500 \text{ g}$$

So q_{water} is

$$q_{water} = (500 \text{ g}) \left(4.18 \, \frac{\text{J}}{\text{g·K}}\right) (297 \text{ K} - 293 \text{ K}) = 8.36 \times 10^3 \text{ J}$$

Setting q_{lead} equal to $-q_{water}$ and solving for $T_{initial}$ yields 473 K:

$$(375 \text{ g}) \left(0.127 \, \frac{\text{J}}{\text{g·K}}\right) \left(297 \text{ K} - T_{initial}\right) = -8.36 \times 10^3 \text{ J}$$
$$297 \text{ K} - T_{initial} = -176 \text{ K}$$
$$T_{initial} = 473 \text{ K}$$

3 **2.3×10^2 g.** To solve this problem, apply $q = mC_p \Delta T$ to the beaker. The unknown is the mass of the beaker, m. Because heat flowed out of the beaker, the sign of q must be negative.

$$-576 \text{ J} = m \left(0.84 \; \frac{\text{J}}{\text{g} \cdot \text{K}} \right) (410 \text{ K} - 413 \text{ K})$$

$$\frac{-576 \text{ J}}{\left(0.84 \; \frac{\text{J}}{\text{g} \cdot \text{K}} \right) (410 \text{ K} - 413 \text{ K})} = m$$

$$2.3 \times 10^2 \text{ g} = m$$

4 **303 K.** Solve this problem in two parts. First, calculate the amount of heat released during dissolution of the NaOH by determining the number of moles of NaOH; then multiply that by the enthalpy change given for NaOH in the problem (44.2 kJ):

$$(25.4 \text{ g NaOH}) \left(\frac{1 \text{ mol NaOH}}{40.0 \text{ g NaOH}} \right) \left(\frac{44.2 \text{ kJ}}{1 \text{ mol NaOH}} \right) = 28.1 \text{ kJ}$$

Because you're given the solution's heat capacity — not its molar heat capacity or its specific heat capacity — you can simply substitute the released heat, q, into the following equation:

$$q = (\text{heat capacity}) \left(T_{\text{final}} - T_{\text{initial}} \right)$$

Because the given heat capacity uses units of joules, you must convert the heat from kilojoules to joules before plugging in the value:

$$2.81 \times 10^4 \text{ J} = \left(4.18 \times 10^3 \; \frac{\text{J}}{\text{K}} \right) \left(T_{\text{final}} - 296 \text{ K} \right)$$

$$\frac{2.81 \times 10^4 \text{ J}}{\left(4.18 \times 10^3 \; \frac{\text{J}}{\text{K}} \right)} = \frac{\left(4.18 \times 10^3 \; \frac{\text{J}}{\text{K}} \right)}{\left(4.18 \times 10^3 \; \frac{\text{J}}{\text{K}} \right)} \left(T_{\text{final}} - 296 \text{ K} \right)$$

$$6.72 \text{ K} = \left(T_{\text{final}} - 296 \text{ K} \right)$$

$$6.72 \text{ K} - (-296) = T_{\text{final}}$$

$$T_{\text{final}} = 303 \text{ K}$$

5 **−83.9 kJ is released.** Solve this problem with a chain of conversion factors. Convert from grams of CO to moles and then from moles to kilojoules. Be sure to adjust the sign of the enthalpy and incorporate the stoichiometry of the given reaction equation. To do so, divide the given mass of CO by the molar mass of CO (28.01 g). You then multiply by the given ΔH and divide by the coefficient of 2 for the CO in the equation.

$$(9.67 \text{ g CO}) \left(\frac{1 \text{ mol CO}}{28.01 \text{ g CO}} \right) \left(\frac{-486 \text{ kJ}}{2 \text{ mol CO}} \right) = -83.9 \text{ kJ}$$

Because the sign is negative, 83.9 kJ is released, not absorbed.

| 6 | **2.60×10^5 J.** The total heat required is the sum of several individual heats: heat to warm the ice to the melting point (273 K), heat to convert the ice to liquid water, heat to warm the liquid water to the boiling point, and heat to convert the liquid water to steam. Be careful to match your units (joules versus kilojoules). To perform each of the conversions, you need to multiply the given mole value by the corresponding molar enthalpies for each phase change. You then use the specific heat formula, $q = mC_p\Delta T$, to calculate the amount of energy needed to raise the temperature of the H_2O between the phase changes. |

Warm ice to melting point: $q = (4.77 \text{ mol}) \left(38.1 \dfrac{\text{J}}{\text{mol} \cdot \text{K}} \right) (273 \text{ K} - 268 \text{ K}) = 909 \text{ J}$

Convert ice to liquid water: $q = (4.77 \text{ mol}) \left(6.01 \times 10^3 \dfrac{\text{J}}{\text{mol}} \right) = 2.87 \times 10^4 \text{ J}$

Warm water to boiling point: $q = (4.77 \text{ mol}) \left(75.3 \dfrac{\text{J}}{\text{mol} \cdot \text{K}} \right) (373 \text{ K} - 273 \text{ K}) = 3.59 \times 10^4 \text{ J}$

Convert liquid water to steam: $q = (4.77 \text{ mol}) \left(4.068 \times 10^4 \dfrac{\text{J}}{\text{mol}} \right) = 1.94 \times 10^5 \text{ J}$

$$q_{\text{total}} = 909 \text{ J} + 2.87 \times 10^4 \text{ J} + 3.59 \times 10^4 \text{ J} + 1.94 \times 10^5 \text{ J} = 2.60 \times 10^5 \text{ J}$$

The sum of all heats is 2.60×10^5 J (or 2.60×10^2 kJ).

| 7 | **−146 kJ.** Reverse Reaction 1 to get Reaction 1′ (−1,032 kJ). Reverse Reaction 2 to get Reaction 2′ (886 kJ). Add Reactions 1′ and 2′, yielding $\Delta H = -146$ kJ. |

Reaction 1′:	$2NO_2(g) \rightarrow N_2(g) + 2O_2(g)$	$\Delta H = -1{,}032$ kJ
Reaction 2′:	$N_2O_4(g) \rightarrow 2NO_2(g)$	$\Delta H = 886$ kJ
sum:	$N_2O_4(g) \rightarrow N_2(g) + 2O_2(g)$	$\Delta H = -146$ kJ

Part IV
Swapping Charges

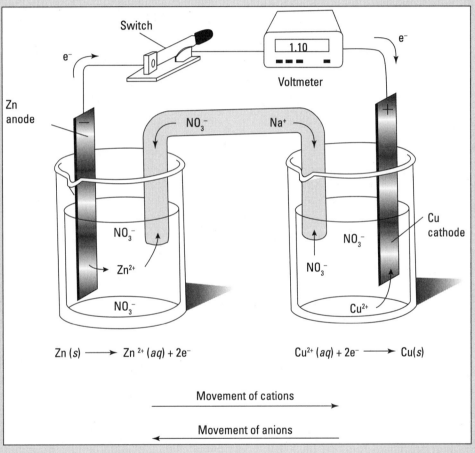

© John Wiley & Sons, Inc.

Find out more about equivalents and normality in an article at www.dummies.com/extras/chemistrywb.

In this part . . .

✔ There are several ways of identifying whether a compound is an acid or a base, depending on what it does with protons and electrons; pH and pOH calculations, along with the values of dissociation constants (K_a and K_b), can help chemists determine the properties of these acids and bases.

✔ Beyond simple pH and pOH lie titrations and buffers. Titrations allow you to determine the concentration of acids and bases. Buffers maintain the pH of a solution by reacting to changes and neutralizing them.

✔ Redox reactions, which involve a transfer of electrons, can occur in acidic and basic conditions. Electrochemistry explains the creation of galvanic and electrolytic cells. You find out about both topics in this part.

✔ Radioactive decay can take place in multiple ways. Chemists can use the type of decay to help determine the half-life of a radioactive isotope.

Chapter 16

Working with Acids and Bases

*I*f you've read any comic books, watched any superhero flicks, or even tuned in to one of those fictional solve-the-crime-in-50-minutes shows on TV, you've likely come across a reference to acids as dangerous substances. Acids are generally thought of as something that evil villains intend to spray in the face of a hero or heroine but somehow usually manage to spill on themselves. However, you encounter and even ingest a wide variety of fairly harmless acids in everyday life. Citric acid, present in citrus fruits such as lemons and oranges, is very ingestible, as is acetic acid, also known as *vinegar.*

Strong acids can indeed burn the skin and must be handled with care in the laboratory. However, strong bases can burn skin as well. Chemists must have a more sophisticated understanding of the differences between an acid and a base and their relative strengths than simply their propensity to burn. This chapter focuses on how you can identify acids and bases as well as several ways to determine their strengths.

Surveying Three Complementary Methods for Defining Acids and Bases

As chemists came to understand acids and bases as more than just "stuff that burns," their understanding of how to define them evolved as well. It's often said that acids taste sour, while bases taste bitter, but we do *not* recommend that you go around tasting chemicals in the laboratory to identify them as acids or bases. In the following sections, we explain three much safer methods you can use to tell the difference between the two.

There's one thing to get out of the way right now, though. As you deal with acids and bases, you see water (H_2O) referred to as an acid at certain points and as a base at others. This is totally fine because water is *amphoteric,* which means it can act as an acid or a base as needed.

Method 1: Arrhenius sticks to the basics

Svante Arrhenius was a Swedish chemist who is credited not only with the acid-base determination method that's named for him but also with an even more fundamental chemical concept: that of *dissociation*. In his PhD thesis, Arrhenius proposed an explanation for a phenomenon that, at the time, had chemists all over the world scratching their heads. What had them perplexed was this: Although neither pure salt nor pure water is a good conductor of electricity, solutions in which salts are dissolved in water tend to be excellent conductors of electricity.

Arrhenius proposed that aqueous solutions of salts conducted electricity because the bonds between atoms in the salts had been broken simply by mixing them into the water, forming ions. Although Michael Faraday had defined the ion several decades earlier, chemists at the time generally believed that ions could form only through *electrolysis,* or the breaking of chemical bonds using electric currents, so Arrhenius's theory was met with some skepticism. Ironically, although his thesis committee wasn't overly impressed and gave him a grade just barely sufficient to pass, Arrhenius eventually managed to win over the scientific community with his research. He was awarded the Nobel Prize in Chemistry in 1903 for the same ideas that nearly cost him his doctorate.

Arrhenius subsequently expanded his theories to form one of the most widely used and straightforward definitions of acids and bases. Arrhenius said that *acids* are substances that form hydrogen (H^+) ions when they dissociate in water, while *bases* are substances that form hydroxide (OH^-) ions when they dissociate in water.

Peruse Table 16-1 for a list of common acids and bases, noting that all the acids in the list contain a hydrogen at the beginnings of their formulas and that most of the bases contain a hydroxide. The Arrhenius definition of acids and bases is straightforward and works for many common acids and bases, but it's limited by its narrow definition of bases.

Table 16-1	Common Acids and Bases		
Acid Name	*Chemical Formula*	*Base Name*	*Chemical Formula*
Acetic acid	$HC_2H_3O_2$ (or CH_3COOH)	Ammonia	NH_3
Citric acid	$HC_6H_7O_7$ (or $C_5H_7O_5COOH$)	Calcium hydroxide	$Ca(OH)_2$
Hydrochloric acid	HCl	Magnesium hydroxide	$Mg(OH)_2$
Hydrofluoric acid	HF	Potassium hydroxide	KOH
Nitric acid	HNO_3	Sodium carbonate	Na_2CO_3
Nitrous acid	HNO_2	Sodium hydroxide	NaOH
Sulfuric acid	H_2SO_4		

Method 2: Brønsted-Lowry tackles bases without a hydroxide ion

You no doubt noticed that some of the bases in Table 16-1 don't contain a hydroxide ion, which means that the Arrhenius definition of acids and bases can't apply. When chemists realized that several substances behaved like bases but didn't contain a hydroxide ion, they reluctantly acknowledged that another determination method was needed. Independently proposed by Johannes Brønsted and Thomas Lowry in 1923 and therefore named after both of them, the Brønsted-Lowry method for determining acids and bases accounts for those pesky non-hydroxide–containing bases.

Under the Brønsted-Lowry definition, an *acid* is a substance that donates a hydrogen ion (H^+) in an acid-base reaction, while a *base* is a substance that accepts that hydrogen ion from the acid. When ionized to form a hydrogen cation, hydrogen loses its one and only electron and is left with only a single proton. For this reason, Brønsted-Lowry acids are often called *proton donors,* and Brønsted-Lowry bases are called *proton acceptors.*

The best way to spot Brønsted-Lowry acids and bases is to keep careful track of hydrogen ions in a chemical equation. Consider, for example, the dissociation of the base sodium carbonate in water. Note that although sodium carbonate is a base, it doesn't contain a hydroxide ion.

$$Na_2CO_3 + 2H_2O \rightarrow H_2CO_3 + 2NaOH$$

This is a simple double replacement reaction (see Chapter 8 for an introduction to these types of reactions). A hydrogen ion from water switches places with the sodium of sodium carbonate to form the products carbonic acid and sodium hydroxide. By the Brønsted-Lowry definition, water is the acid because it donates its hydrogen to Na_2CO_3. This makes Na_2CO_3 the base because it accepts the hydrogen from H_2O.

What about the substances on the right-hand side of the equation? Brønsted-Lowry theory calls the products of an acid-base reaction the *conjugate acid* and *conjugate base.* The conjugate acid (in this case, H_2CO_3) is produced when the base accepts a proton, while the conjugate base (NaOH) is formed when the acid loses its hydrogen. This reaction also brings up a very important point about the strength of each of these acids and bases. Although sodium carbonate is a very strong base, its conjugate acid, carbonic acid, is a very weak acid. Similarly, water is an extremely weak acid, and its conjugate base, sodium hydroxide, is a very strong base. Weak acids always form strong conjugate bases and vice versa. The same is true of strong acids.

Method 3: Lewis relies on electron pairs

In the same year that Brønsted and Lowry proposed their definition of acids and bases, an American chemist named Gilbert Lewis proposed an alternative definition that not only encompassed Brønsted-Lowry theory but also accounted for acid-base reactions in which a hydrogen ion isn't exchanged. Lewis's definition relies on tracking lone pairs of electrons. Under his theory, a *base* is any substance that donates a pair of electrons to form a coordinate covalent bond with another substance, while an *acid* is a substance that accepts that electron pair in such a reaction. As we explain in Chapter 5, a *coordinate covalent bond* is a covalent bond in which both of the bonding electrons are donated by one of the atoms forming the bond.

All Brønsted-Lowry acids are Lewis acids, but in practice, the term *Lewis acid* is generally reserved for Lewis acids that don't also fit the Brønsted-Lowry definition. The best way to spot a Lewis acid-base pair is to draw a Lewis dot structure of the reacting substances, noting the presence of lone pairs of electrons. (We introduce Lewis structures in Chapter 5.) For example, consider the reaction between ammonia (NH_3) and boron trifluoride (BF_3):

$$NH_3 + BF_3 \rightarrow NH_3BF_3$$

At first glance, neither the reactants nor the product appears to be an acid or base, but the reactants are revealed as a Lewis acid-base pair when drawn as Lewis dot structures as in Figure 16-1. Ammonia donates its lone pair of electrons to the bond with boron trifluoride, making ammonia the Lewis base and boron trifluoride the Lewis acid.

Figure 16-1:
The Lewis dot structures of ammonia and boron trifluoride.

© John Wiley & Sons, Inc.

Sometimes you can identify the Lewis acid and base in a compound without drawing the Lewis dot structure. You can do this by identifying reactants that are electron rich (bases) or electron poor (acids). A metal cation, for example, is electron poor and tends to act as a Lewis acid in a reaction, accepting a pair of electrons.

TIP

In practice, it's much simpler to use the Arrhenius or Brønsted-Lowry definition of acid and base, but you'll need to use the Lewis definition when hydrogen ions aren't being exchanged. You can pick and choose among the definitions when you're asked to identify the acid and base in a reaction.

EXAMPLE

Q. Identify the acid and base in the following reaction and label their conjugates.

$$NH_3 + H_2O \rightarrow NH_4^+ + OH^-$$

A. This is a classic Brønsted-Lowry acid-base pair. Water (H_2O) loses a proton to ammonia (NH_3), forming a hydroxide anion. This makes water the proton donor, or Brønsted-Lowry acid, and OH^- its conjugate base. Ammonia accepts the proton from water, forming ammonium (NH_4^+). This makes ammonia the proton acceptor, or Brønsted-Lowry base, and NH_4^+ its conjugate acid.

1. Consider the following reaction. Label the acid, base, conjugate acid, and conjugate base, and comment on their strengths. How can water act as an acid in one reaction and a base in another?

$$HCl + H_2O \rightarrow H_3O^+ + Cl^+$$

Solve It

2. Use the Arrhenius definition to identify the acid or base in each reaction and explain how you know.

a. $NaOH(s) + H_2O \rightarrow Na^+(aq) + OH^-(aq) + H_2O$

b. $HF(g) + H_2O \rightarrow H^+(aq) + F^-(aq) + H_2O$

Solve It

3. Identify the Lewis acid and base in each reaction. Draw Lewis dot structures for the first two, and determine the Lewis acid and base in the third reaction without a dot structure.

a. $6H_2O + Cr^{3+} \rightarrow Cr(OH_2)_6^{3+}$

b. $2NH_3 + Ag^+ \rightarrow Ag(NH_3)_2^+$

c. $2Cl^- + HgCl_2 \rightarrow HgCl_4^{2-}$

Solve It

Measuring Acidity and Basicity: pH, pOH, and K_W

A substance's identity as an acid or a base is only one of many things that a chemist may need to know about it. Sulfuric acid and water, for example, can both act as acids, but using sulfuric acid to wash your face in the morning would be a grave error indeed. Sulfuric acid and water differ greatly in *acidity,* a measurement of an acid's strength. A similar quantity, called *basicity,* measures a base's strength.

Acidity and basicity are measured in terms of quantities called *pH* and *pOH,* respectively. Both are simple scales ranging from 0 to 14, with low numbers on the pH scale representing a higher acidity and therefore a stronger acid. On both scales, a measurement of 7 indicates a *neutral solution.* On the pH scale, any number lower than 7 indicates that the solution is acidic, with acidity increasing as pH decreases, and any number higher than 7 indicates a basic solution, with basicity increasing as pH increases. In other words, the further the pH gets away from 7, the more acidic or basic a substance gets. The pOH shows exactly the same relationship between distance from 7 and acidity or basicity, only this time, low numbers indicate very basic solutions, while high numbers indicate very acidic solutions.

You calculate pH using the formula $pH = -\log[H^+]$, where the brackets around H+ indicate that it's a measurement of the concentration of hydrogen ions in moles per liter (or molarity; see Chapter 12). You calculate pOH using a similar formula, with OH⁻ concentration replacing the H+ concentration: $pOH = -\log[OH^-]$. The word *log* in each formula stands for logarithm.

Because a substance with high acidity must have low basicity, a low pH indicates a high pOH for a substance and vice versa. In fact, a very convenient relationship between pH and pOH allows you to solve for one when you have the other: $pH + pOH = 14$.

You'll often be given a pH or pOH and be asked to solve for the H+ or OH⁻ concentrations instead of the other way around. The logarithms in the pH and pOH equations make it tricky to solve for [H+] or [OH⁻], but if you remember that a log is undone by raising 10 to both sides of an equation, you quickly arrive at a convenient formula for [H+], namely $[H^+] = 10^{-pH}$. Similarly, [OH⁻] can be calculated using the formula $[OH^-] = 10^{-pOH}$. As with pH and pOH, a convenient relationship exists between [H+] and [OH⁻], which multiply together to equal a constant. This constant, called the *ion product constant for water,* or K_w, is calculated as follows:

$$K_W = [H^+][OH^-] = 1 \times 10^{-14}$$

Q. Calculate the pH and pOH of a solution with an [H⁺] of 1×10^{-8}. Is the solution acidic or basic? Do the same for a solution with an [OH⁻] of 2.3×10^{-11}.

A. For a solution with an [H⁺] of 1×10^{-8}: **pH = 8; pOH = 6; the solution is a base.** You've been given the H⁺ concentration, so first solve for the pH by plugging [H⁺] into the formula for pH:

$$pH = -\log\left[1\times10^{-8}\right] = 8$$

Plugging this value into the equation relating pH and pOH gives you a value for pOH:

$$8 + pOH = 14$$
$$pOH = 14 - 8$$
$$pOH = 6$$

A pH of 8 indicates that this solution is very slightly basic. It's not just a coincidence that the exponent of the H⁺ concentration is equal to the pH. This is true whenever the coefficient of the H⁺ concentration is 1.

For a solution with an [OH⁻] of 2.3×10^{-11}: **pOH = 10.6; pH = 3.4; the solution is an acid.** The second portion of the problem gives you an [OH⁻], in which case you can use the ion-product constant for water to calculate the [H⁺] and then solve for pH and pOH as you did before; or more simply, you can first calculate the pOH and use it to find the pH. Plugging an [OH⁻] of 2.3×10^{-11} into the pOH equation yields

$$pOH = -\log\left[2.3\times10^{-11}\right] = 10.6$$

Plug this value into the relation between pH and pOH to get the pH:

$$pH + 10.6 = 14$$
$$pH = 14 - 10.6$$
$$pH = 3.4$$

This low pH indicates that the substance is a relatively strong acid.

---

4. Determine the pH given the following values:

a. $[H^+] = 1\times10^{-13}$

b. $[H^+] = 1.58\times10^{-9}$

c. $[OH^-] = 2.7\times10^{-7}$

d. $[OH^-] = 1\times10^{-7}$

Solve It

5. Determine the pOH given the following values:

a. $[H^+] = 2\times10^{-8}$

b. $[OH^-] = 5.1\times10^{-11}$

Solve It

6. Determine whether the following are acidic, basic, or neutral:

a. $[OH^-] = 2.5 \times 10^{-7}$

b. $[OH^-] = 3.1 \times 10^{-13}$

c. $[H^+] = 4.21 \times 10^{-5}$

d. $[H^+] = 8.9 \times 10^{-10}$

Solve It

7. Determine the [H⁺] from the following pH or pOH values:

a. pH = 3.3

b. pH = 7.69

c. pOH = 10.21

d. pOH = 1.26

Solve It

K_a and K_b: Finding Strength through Dissociation

Arrhenius's concept of dissociation (which we cover earlier in this chapter) gives you a convenient way of measuring the strength of an acid or base. Although water tends to dissociate all acids and bases, the degree to which they dissociate depends on their strength. Strong acids such as HCl, HNO_3, and H_2SO_4 dissociate completely in water, while weak acids dissociate only partially. Practically speaking, a weak acid is any acid that doesn't dissociate completely in water.

See Table 16-2 for a list of strong acids and strong bases.

Table 16-2		Strong Acids and Bases	
Acid Name	*Chemical Formula*	*Base Name*	*Chemical Formula*
Hydrochloric acid	HCl	Lithium hydroxide	LiOH
Hydroiodic acid	HI	Sodium hydroxide	NaOH
Hydrobromic acid	HBr	Potassium hydroxide	KOH
Nitric acid	HNO_3	Rubidium hydroxide	RbOH
Sulfuric acid	H_2SO_4	Cesium hydroxide	CsOH
Chloric acid	$HClO_3$	Calcium hydroxide	$Ca(OH)_2$
Perchloric acid	$HClO_4$	Strontium hydroxide	$Sr(OH)_2$
		Barium hydroxide	$Ba(OH)_2$

To measure the amount of dissociation occurring when a weak acid is in aqueous solution, chemists use a constant called the *acid dissociation constant* (K_a). K_a is a special variety of the equilibrium constant. As we explain in Chapter 14, the equilibrium constant of a chemical reaction is the concentration of products over the concentration of reactants, and it indicates the balance between products and reactants in a reaction.

The *acid dissociation constant* is simply the equilibrium constant of a reaction in which an acid is mixed with water and from which the water concentration has been removed. The water concentration is removed because the concentration of water is a constant in dilute solutions, and a better indicator of acidity is the concentration of the dissociated products divided by the concentration of the acid reactant. The general form of the acid dissociation constant is therefore

$$K_a = \frac{[H_3O^+][A^-]}{[HA]}$$

where [HA] is the concentration of the acid before it loses its hydrogen and [A⁻] is the concentration of its conjugate base. Notice that the concentration of the hydronium ion (H_3O^+) is used in place of the concentration of H^+, which we use to describe acids earlier in this chapter. In truth, they're one and the same. Generally speaking, H^+ ions in aqueous solution will be caught up by atoms of water in solution, making H_3O^+ ions.

A similar situation exists for bases. Strong bases such as KOH, NaOH, and $Ca(OH)_2$ dissociate completely in water. Weak bases don't dissociate completely in water, and their strength is measured by the *base dissociation constant, K_b*:

$$K_b = \frac{[OH^-][B^+]}{[BOH]}$$

Here, BOH is the base, and B⁺ is its conjugate acid. You can also write K_b in terms of the acid and base:

$$K_b = \frac{[OH^-][HA]}{[A^-]}$$

In problems where you're asked to calculate K_a or K_b, you'll generally be given the concentration (molarity) of the original acid or base and the concentration of either its conjugate *or* hydronium/hydroxide. Dissociation of a single molecule of acid involves the splitting of that acid into one molecule of its conjugate base and one hydrogen ion, and dissociation of a base always involves the splitting of that base into one molecule of the conjugate acid and one hydroxide ion. For this reason, the concentration of the conjugate and the concentration of the hydronium or hydroxide ion in any dissociation are equal, so you need only one to know the other.

You may be given the pH and be asked to figure out the [H⁺] (equivalent to [H_3O^+]) or [OH⁻]. K_a and K_b are also constants at constant temperature.

Q. Write a general expression for the acid dissociation constant of the following reaction, a dissociation of ethanoic acid:

$$CH_3COOH + H_2O \rightarrow H_3O^+ + CH_3COO^-$$

Then calculate its actual value if $[CH_3COOH] = 2.34 \times 10^{-4}$ and $[CH_3COO^-] = 6.51 \times 10^{-5}$.

A. 1.80×10^{-5}. Your general expression should be

$$K_a = \frac{[H_3O^+][CH_3COO^-]}{[CH_3COOH]}$$

You're given $[CH_3COO^-] = 6.51 \times 10^{-5}$, and $[H_3O^+]$ must be the same. Plugging all known values into the expression for K_a yields

$$K_a = \frac{[6.51 \times 10^{-5}][6.51 \times 10^{-5}]}{[2.34 \times 10^{-4}]} = 1.80 \times 10^{-5}$$

The K_a is very small, so this is a very weak acid.

---

8. Calculate the K_a for a 0.50 M solution of benzoic acid, C_6H_5COOH, if the $[H^+]$ concentration is 5.6×10^{-3}.

Solve It

9. The pH of a 0.75 M solution of HCOOH is 1.93. Calculate the acid dissociation constant.

Solve It

---

10. If you have a 0.20 M solution of ammonia, NH_3, that has a K_b of 1.80×10^{-5}, what is the pH of that solution?

Solve It

Answers to Questions on Acids and Bases

The following are the answers to the practice problems presented in this chapter.

1 In this reaction, hydrochloric acid (HCl) donates a proton to water (H_2O), making it the Brønsted-Lowry acid. Water, which accepts the proton, is the Brønsted-Lowry base. This makes hydronium (H_3O^+) the conjugate acid and chloride (Cl^-) the conjugate base. Water can act as the base in this reaction and as an acid in the example problem because it's composed of both a hydrogen ion and a hydroxide ion; therefore, it can either accept or donate a proton.

2 For Arrhenius acids, remember to track the movement of H^+ and OH^- ions. If the reaction yields an OH^- product, the substance is a base, whereas an H^+ product reveals that the substance is an acid.

   a. NaOH, Arrhenius base. Sodium hydroxide (NaOH) dissociates in water to form OH^- ions, making it an Arrhenius base.

   b. HF, Arrhenius acid. Hydrofluoric acid (HF) dissociates in aqueous solution to form H^+ ions, making it an Arrhenius acid.

3 To identify Lewis acids and bases, track the movement of electron pairs. Draw a Lewis dot structure to locate the atom with a lone pair available to donate. This is the Lewis base.

   a. Water (H_2O) is the Lewis base, and chromium (Cr^{3+}) is the Lewis acid. Your Lewis dot structure should show that the lone pair of electrons is donated to the bond by H_2O. You can tell that the electrons come from H_2O because chromium doesn't have any electrons available for bonding due to its ionic charge of +3.

© John Wiley & Sons, Inc.

   b. Ammonia (NH_3) is the Lewis base, and silver (Ag^+) is the Lewis acid. Your Lewis dot structure should show that the lone pair of electrons is donated to the bond by NH_3. The positive metal (silver) doesn't have any electrons available for bonding, so the electrons for forming the bond must come from NH_3.

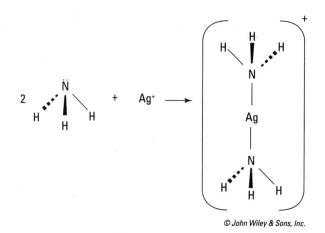

© John Wiley & Sons, Inc.

c. **The negative charge on the chlorine (Cl⁻) indicates that the chlorine is electron rich, making it the Lewis base. Mercury (HgCl₂) is therefore the Lewis acid.** Mercury is a metal and regularly forms a positive ion, which is another hint that mercury will act as the Lewis acid and accept the electrons.

4 To calculate the pH values in this problem, use the equation $pH = -\log[H^+]$ and plug in all known values. If you're given [OH⁻] instead of [H⁺], simply calculate the pOH (using the equation $pOH = -\log[OH^-]$) instead and then subtract that number from 14 to give you the pH.

a. pH = 13. Because the coefficient on the H⁺ concentration is 1, the pH is simply the exponent.

b. pH = 8.8. Plug the given [H⁺] into the pH equation, giving you $pH = -\log[1.58 \times 10^{-9}] = 8.8$.

c. pH = 7.4. First, use the OH⁻ concentration to calculate the pOH: $pOH = -\log[2.7 \times 10^{-7}] = 6.6$. Then use the relationship between pH and pOH ($14 - pOH = pH$) to determine the pH: $14 - 6.6 = 7.4$.

d. pH = 7. Because the coefficient of this OH⁻ concentration is 1, you can read the pOH directly from the exponent to get pOH = 7. Then use the relationship between pH and pOH ($14 - pOH = pH$) to determine the pH: $14 - 7 = 7$.

5 To calculate pOH, simply take the negative logarithm of the OH⁻ concentration. If you're given the H⁺ concentration, use it to calculate the pH and then subtract that value from 14 to yield the pOH.

a. pOH = 6.3. First, use the H⁺ concentration to calculate the pH: $pH = -\log[2 \times 10^{-8}] = 7.7$. Then use the relationship between pH and pOH ($14 - pH = pOH$) to determine the pOH: $14 - 7.7 = 6.3$.

b. pOH = 10.3. Here, you can simply use the OH⁻ concentration to calculate the pOH directly: $pOH = -\log[5.1 \times 10^{-11}] = 10.3$.

6 To determine acidity, you need to calculate pH.

a. Basic. Use [OH⁻] to determine pOH: $pOH = -\log[2.5 \times 10^{-7}] = 6.6$. Subtract this value from 14 to get a pH of 7.4. This pH is very slightly greater than 7, so the solution is slightly basic.

b. Acidic. Use [OH⁻] to determine pOH: $pOH = -\log[3.1 \times 10^{-13}] = 12.5$. Subtract this value from 14 to get a pH of 1.5. This pH is significantly smaller than 7, so the solution is quite acidic.

c. **Acidic.** Use the [H⁺] concentration to determine pH: $pH = -\log\left[4.21 \times 10^{-5}\right] = 4.4$. This solution has a pH smaller than 7 and is therefore acidic.

d. **Basic.** Use the [H⁺] concentration to determine pH: $pH = -\log\left[8.9 \times 10^{-10}\right] = 9.1$. This solution has a pH greater than 7 and is therefore basic.

7 In this problem, use the formula $\left[H^+\right] = 10^{-pH}$. If you're given the pOH instead of the pH, begin by subtracting the pOH from 14 to give you the pH and then plug that number into the formula.

a. **5.0×10^{-4}.** Use the pH to calculate the H⁺ concentration using the equation $\left[H^+\right] = 10^{-pH}$: $\left[H^+\right] = 10^{-3.3} = 5.0 \times 10^{-4}$.

b. **2.04×10^{-8}.** Use the pH to calculate the H⁺ concentration using the equation $\left[H^+\right] = 10^{-pH}$: $\left[H^+\right] = 10^{-7.69} = 2.04 \times 10^{-8}$.

c. **1.62×10^{-4}.** Begin by using the pOH to calculate the pH using the equation $14 - pOH = pH$: $14 - 10.21 = 3.79$. Then use the pH to calculate the H⁺ concentration using the equation $\left[H^+\right] = 10^{-pH}$: $\left[H^+\right] = 10^{-3.79} = 1.62 \times 10^{-4}$.

d. **1.82×10^{-13}.** Begin by using the pOH to calculate the pH using the equation $14 - pOH = pH$: $14 - 1.26 = 12.74$. Then use the pH to calculate the H⁺ concentration using the equation $\left[H^+\right] = 10^{-pH}$: $\left[H^+\right] = 10^{-12.74} = 1.82 \times 10^{-13}$.

8 **6.3×10^{-5}.** The molarity in this problem tells you that the concentration of benzoic acid is 5.0×10^{-1}. You also know that the concentration of benzoic acid's conjugate base is the same as the given H⁺ concentration. All that remains is to write an equation for the acid dissociation constant and plug in these concentrations.

$$K_a = \frac{\left[H_3O^+\right]\left[C_6H_5COO^-\right]}{\left[C_6H_5COOH\right]} = \frac{\left[5.6 \times 10^{-3}\right]\left[5.6 \times 10^{-3}\right]}{\left[5.0 \times 10^{-1}\right]} = 6.3 \times 10^{-5}$$

9 **1.6×10^{-4}.** You need to use the pH to calculate the [H⁺] with $\left[H^+\right] = 10^{-pH}$:

$$\left[H^+\right] = 10^{-1.93} = 1.1 \times 10^{-2}$$

This value must also be equal to the concentration of the conjugate base. All that remains is to write an equation for the acid dissociation constant and plug in these concentrations:

$$K_a = \frac{\left[H_3O^+\right]\left[HCOO^-\right]}{\left[HCOOH\right]} = \frac{\left[1.1 \times 10^{-2}\right]\left[1.1 \times 10^{-2}\right]}{\left[7.5 \times 10^{-1}\right]} = 1.6 \times 10^{-4}$$

10 **pH = 11.3.** Begin by considering the K_b equation:

$$K_b = \frac{\left[OH^-\right]\left[NH_4^+\right]}{\left[NH_3\right]}$$

Because $\left[OH^-\right] = \left[NH_4^+\right]$, you can rewrite this as

$$K_b = \frac{\left[OH^-\right]^2}{\left[NH_3\right]}$$

Solving this equation for [OH⁻] yields $[OH^-] = \sqrt{(K_b)([NH_3])}$. Plugging in the known values of K_b and $[NH_3]$ into this equation yields

$$[OH^-] = \sqrt{(1.8\times10^{-5})([0.20])} = 1.9\times10^{-3}$$

Use this to solve for the pOH:

$$pOH = -\log[1.9\times10^{-3}] = 2.7$$

The final step is to solve for pH by subtracting this number from 14:

$$14 - 2.7 = 11.3$$

Chapter 17

Achieving Neutrality with Titrations and Buffers

. . . .

In This Chapter

▶ Deducing the concentration of a mystery acid or base through titration

▶ Working with buffered solutions

▶ Measuring the solubility of saturated solutions of salts with K_{sp}

. . . .

*I*n the real world of chemistry, acids and bases often meet in solution, and when they do, they're drawn to one another. These unions of acid and base are called *neutralization reactions* because the lower pH of the acid and the higher pH of the base essentially cancel one another out, resulting in a neutral solution. In fact, when a hydroxide-containing base reacts with an acid (containing H^+), the products are simply an innocuous salt and water.

Although strong acids and bases have their uses, the prolonged presence of a strong acid or base in an environment not equipped to handle it can be very damaging. In the laboratory, for example, you need to handle strong acids and bases carefully, deliberately performing neutralization reactions when appropriate. A lazy chemistry student who attempts to dump a concentrated acid or base down the laboratory sink will get a browbeating from her hawk-eyed teacher. Dumping stuff like this can damage the pipes and is generally unsafe. A responsible chemist neutralizes acidic laboratory waste with a base such as baking soda and neutralizes basic waste with an acid. Doing so makes a solution perfectly safe to dump down the drain and often results in the creation of a satisfyingly sizzly solution while the reaction is occurring.

In this chapter, we explore three issues that arise when acids meet up with bases:

✔ **Titration:** Titrations are the process by which chemists add acids to bases (or vice versa) a little bit at a time, gradually using up acid and base equivalents as the two neutralize each other. We show you how to use titration to figure out the concentration of an unknown acid.

✔ **Buffer solutions:** Buffer solutions are mixtures that contain both acid and base forms of the same compounds and serve to maintain the pH of the solution even when extra acid or base is added.

✔ **Solubility product:** Because salts are produced when acids react with bases, we discuss the solubility product, K_{sp}, a number that tells you how soluble a salt is in solution.

At heart, neutralization reactions in which the base contains a hydroxide ion are simple double-replacement reactions of the form $HA + BOH \rightarrow BA + H_2O$ (in other words, an acid reacts with a base to form a salt and water). You're asked to write a number of such reactions in this chapter, so be sure to review double replacement reactions and balancing equations in Chapter 8 before you delve into the new and exciting world of neutralization.

Concentrating on Titration to Figure Out Molarity

Imagine you're a newly hired laboratory assistant who's been asked to alphabetize the chemicals on the shelves of a chemistry laboratory during a lull in experimenting. As you reach for the bottle of sulfuric acid, your first-day jitters get the better of you, and you knock over the bottle. Some careless chemist failed to screw the cap on tightly! You quickly neutralize the acid with a splash of baking soda and wipe up the now nicely neutral solution. As you pick up the bottle, however, you notice that the spilled acid burned away most of the label! You know it's sulfuric acid, but there are several different concentrations of sulfuric acid on the shelves, and you don't know the molarity of the solution in this bottle. Knowing that your boss will surely blame you if she sees the damaged bottle and not wanting to get sacked on your very first day, you quickly come up with a way to determine the molarity of the solution and save your job.

You know that the bottle contains sulfuric acid of a mystery concentration, and you notice bottles of 1 M sodium hydroxide, a strong base, and phenolphthalein, a pH indicator, among the chemicals on the shelves. You measure a small amount of the mystery acid into a beaker and add a little phenolphthalein. You reason that if you drop small amounts of sodium hydroxide into the solution until the phenolphthalein indicates that the solution is neutral by turning the appropriate color, you'll be able to figure out the acid's concentration.

You can find the concentration by doing a simple calculation of the number of moles of sodium hydroxide you've added and then reasoning that the mystery acid must have an equal number of moles to have been neutralized. This then leads to the number of moles of acid, and that, in turn, can be divided by the volume of acid you added to the beaker to get the molarity. Whew! You relabel the bottle and rejoice in the fact that you can come in and do menial labor in the lab again tomorrow.

 This process is called a *titration,* and it's often used by chemists to determine the molarity of acids and bases. In a titration calculation, you generally know the identity of an acid or base of unknown concentration, and you know the identity and molarity of the acid or base that you're going to use to neutralize it. Given this information, you then follow six simple steps:

1. **Measure out a small volume of the mystery acid or base.**

2. **Add a pH indicator such as phenolphthalein to the mystery acid or base.**

    Different pH indicators involve different color changes. Make sure you know which color the solution will turn when it reaches the equivalence point.

3. Neutralize.

Slowly add the acid or base of known concentration to the solution until the indicator shows that the solution is neutral, keeping careful track of the volume added.

4. Calculate the number of moles added.

Multiply the number of liters of acid or base added by the molarity of that acid or base to get the number of moles you added.

5. Calculate the unknown moles.

Use the balanced equation and basic stoichiometry to determine how many moles of the mystery substance being neutralized are present (see Chapter 9 for details on stoichiometry).

6. Solve for molarity.

Divide the number of moles of the mystery acid or base by the number of liters measured out in Step 1, giving you the molarity.

People often visualize the titration process using a graph that shows the concentration of base on one axis and the pH on the other, as in Figure 17-1. The interaction of the two concentrations traces out a *titration curve,* which has a characteristic *s* shape. At the equivalence point in Figure 17-1, the amount of base present is equal to the amount of acid present in the solution. If you're using an indicator such as phenolphthalein, the equivalence point marks when the first permanent color change takes place.

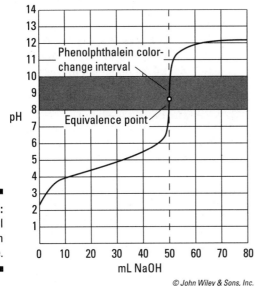

Figure 17-1:
A typical titration curve.

© John Wiley & Sons, Inc.

Q. If the laboratory assistant had to add 10 mL of 1 M sodium hydroxide to neutralize 5 mL of the sulfuric acid in his impromptu titration, what did he end up writing on the label for the concentration of the sulfuric acid in the bottle?

A. **1 M H_2SO_4.** The problem tells you that the volume from Step 1 of the titration process is 5 mL and that the volume of base from Step 3 is 10 mL. In Step 4, you must calculate the number of moles of sodium hydroxide the lab assistant added by multiplying the volume in liters (0.01 L) by the molarity (1 M) to give you 0.01 mol NaOH. The balanced equation for this reaction is

$$H_2SO_4 + 2NaOH \rightarrow 2H_2O + Na_2SO_4$$

You use the coefficients from this equation along with the mole value calculated in Step 4 to determine the mole value of the unknown substance you're titrating:

$$\left(\frac{0.01 \text{ mol NaOH}}{1} \right) \left(\frac{1 \text{ mol } H_2SO_4}{2 \text{ mol NaOH}} \right) = 0.005 \text{ mol } H_2SO_4$$

The final step is to divide this value by the liters of acid added to get the molarity. Your initial acid volume was given as 5 mL, which converts to 0.005 L.

$$\frac{0.005 \text{ mol } H_2SO_4}{0.005 \text{ L}} = 1 \text{ M } H_2SO_4$$

The assistant should have labeled the bottle as 1 M H_2SO_4.

---

1. In doing a titration of a solution of calcium hydroxide of unknown concentration, a student adds 12 mL of 2 M HCl (hydrochloric acid) and finds that the molarity of the base is 1.25 M. How much $Ca(OH)_2$ must the student have measured out at the start of the titration?

Solve It

2. Titration shows that a 5.0 mL sample of nitrous acid, HNO_2, has a molarity of 0.50. If 8.0 mL of magnesium hydroxide, $Mg(OH)_2$, was added to the acid to accomplish the neutralization, what must the molarity of the base have been?

Solve It

3. How much could a chemist find out about a mystery acid or base through titration if neither its identity nor its concentration is known?

Solve It

Maintaining Your pH with Buffers

You may have noticed that the titration curve shown in Figure 17-1 has a flattened area in the middle where pH doesn't change significantly as base is added. This region is called a *buffer region.*

Certain solutions, called *buffered solutions,* resist changes in pH like a stubborn child resists eating her Brussels sprouts: steadfastly at first but choking them down reluctantly if enough pressure is applied (such as the threat of no dessert). Although buffered solutions maintain their pH very well when relatively small amounts of acid or base are added to them or the solution is diluted, they can withstand the addition of only a certain amount of acid or base before becoming overwhelmed.

Buffers are most often made up of a weak acid and its conjugate base, though they can also be made of a weak base and its conjugate acid. (Conjugate bases and acids are the products in acid-base reactions; see Chapter 16 for details.) A weak acid in aqueous solution will be partially dissociated, and the amount of dissociation depends on its pK_a value (the negative logarithm of its acid dissociation constant). The dissociation will be of the form $HA + H_2O \rightarrow H_3O^+ + A^-$, where A^- is the conjugate base of the acid HA. The acidic proton is taken up by a water molecule, forming hydronium. If HA were a strong acid, approximately 100 percent of the acid would become H_3O^+ and A^-, but because it's a weak acid, only a fraction of the HA dissociates and the rest remains HA.

The K_a of this reaction is defined by

$$K_a = \frac{[H_3O^+][A^-]}{[HA]}$$

Solving this equation for the $[H_3O^+]$ concentration allows you to devise a relationship between the $[H_3O^+]$ and the K_a of a buffer:

$$[H_3O^+] = (K_a)\left(\frac{[HA]}{[A^-]}\right)$$

Taking the negative logarithm of both sides of the equation and manipulating logarithm rules yields an equation called the *Henderson-Hasselbalch equation,* which relates the pH and the pK_a:

$$pH = pK_a + \log\left(\frac{[A^-]}{[HA]}\right)$$

With logarithm rules, you can manipulate this equation to get $\dfrac{[A^-]}{[HA]} = 10^{(pH-pK_a)}$, which may be more useful in certain situations.

The very best buffers and those best able to withstand the addition of both acid and base are those for which [HA] and [A⁻] are approximately equal. When this occurs, the logarithmic term in the Henderson-Hasselbalch equation disappears, and the equation becomes pH = pK_a. When creating a buffered solution, chemists therefore choose an acid that has a pK_a close to the desired pH.

If you add a strong base such as sodium hydroxide (NaOH) to this mixture of dissociated base (A⁻) and undissociated acid (HA), the base's hydroxide is absorbed by the acidic proton, replacing the exceptionally strong base OH⁻ with a relatively weak base A⁻ and minimizing the change in pH:

$$HA + OH \rightarrow H_2O + A^-$$

This causes a slight excess of base in the reaction, but it doesn't affect pH significantly. You can think of the undissociated acid as a reservoir of protons that are available to neutralize any strong base that may be introduced to the solution. As we explain in Chapter 14, when a product is added to a reaction, the equilibrium in the reaction changes to favor the reactants or to "undo" the change in conditions. Because this reaction generates A⁻, the acid dissociation reaction happens less frequently as a result, further stabilizing the pH.

When a strong acid, such as hydrochloric acid (HCl), is added to the mixture, its acidic proton is taken up by the base A⁻, forming HA:

$$H + A^- \rightarrow HA$$

This causes a slight excess of acid in the reaction but doesn't affect pH significantly. It also shifts the balance in the acid dissociation reaction in favor of the products, causing it to happen more frequently and recreating the base A⁻.

Figure 17-2 summarizes the addition of acid and base and their effect on the ratio of products and reactants.

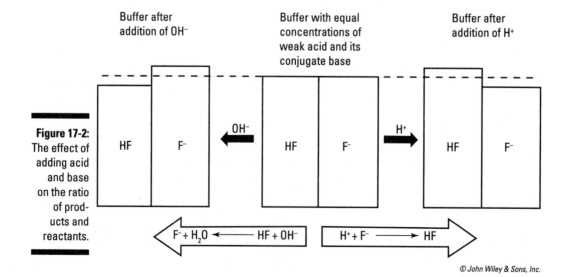

Figure 17-2: The effect of adding acid and base on the ratio of products and reactants.

© John Wiley & Sons, Inc.

Buffers have their limits, however. The acid's proton reservoir, for example, can compensate for the addition of only a certain amount of base before it runs out of protons that can neutralize free hydroxide. At this point, a buffer has done all it can do, and the titration curve resumes its steep upward slope.

Q. Consider a buffered solution that contains the weak acid ethanoic acid ($K_a = 1.8 \times 10^{-5}$) and its conjugate base, ethanoate. If the concentration of the solution is 0.5 M with respect to ethanoic acid and 0.3 M with respect to ethanoate, what is the pH of the solution?

A. **pH = 4.5.** This problem is a simple application of the Henderson-Hasselbalch equation. Remember to take the negative logarithm of K_a to get pK_a. Plugging in known values yields

$$pH = pK_a + \log\left(\frac{[A^-]}{[HA]}\right) = -\log\left(1.8 \times 10^{-5}\right) + \log\left(\frac{[0.3]}{[0.5]}\right) = 4.5$$

---

4. Ethanoic acid would make an ideal buffer to maintain what approximate pH?

Solve It

5. Describe the preparation of 1,000 mL of a 0.2 M carbonic acid buffer of pH = 7.0. Assume the pK_a of carbonic acid is 6.8.

Solve It

Measuring Salt Solubility with K_{sp}

In chemistry, a salt is not necessarily the substance you sprinkle on french fries. Rather, a *salt* is any substance that is a combination of an anion and a cation and is created in a neutralization reaction. Salts, therefore, tend to dissociate in water. The degree of dissociation possible — in other words, the solubility of the salt — varies greatly from one salt to another.

Chemists use a quantity called the *solubility product constant*, or K_{sp}, to compare the solubilities of salts. K_{sp} is calculated in much the same way as an equilibrium constant (K_{eq}; see Chapter 14). The product concentrations are multiplied together, each raised to the power of its coefficient in the balanced dissociation equation. There's one key difference, however, between a K_{sp} and a K_{eq}. K_{sp} is a quantity specific to a *saturated* solution of salt, so the concentration of the undissociated salt reactant has absolutely no bearing on its value. If the solution is saturated, then the amount of possible dissociation is at its maximum, and any additional solute added merely settles on the bottom.

Q. Write a formula for the solubility product constant of the reaction $CaF_2 \rightarrow Ca^{2+} + 2F^-$.

A. $K_{sp} = [Ca^{2+}][F^-]^2$. You construct the solubility product constant by raising the concentrations of the two products to the power of their coefficients, so $K_{sp} = [Ca^{2+}][F^-]^2$.

6. Write the solubility product dissociation constants for silver (I) chromate (Ag_2CrO_4) and strontium sulfate ($SrSO_4$).

Solve It

7. If the K_{sp} of silver (I) chromate is 1.1×10^{-12} and the silver ion concentration in the solution is 0.0005 M, what is the chromate concentration?

Solve It

Answers to Questions on Titrations and Buffers

Following are the answers to the practice problems presented in this chapter.

1   **9.6 mL Ca(OH)$_2$.** Begin by finding the number of moles of HCl by multiplying the molarity (2 M) by the volume (0.012 L) to get 0.024 mol HCl. Next, calculate the number of moles of Ca(OH)$_2$ needed to neutralize this amount using the balanced chemical equation
$Ca(OH)_2 + 2HCl \rightarrow CaCl_2 + 2H_2O$:

$$\left(\frac{0.024 \text{ mol HCl}}{1} \right) \left(\frac{1 \text{ mol Ca(OH)}_2}{2 \text{ mol HCl}} \right) = 0.012 \text{ mol Ca(OH)}_2$$

Divide this number by the molarity of the base to get the volume of base added:

$$\left(\frac{0.012 \text{ mol Ca(OH)}_2}{1.25 \text{ M Ca(OH)}_2} \right) = 0.0096 \text{ L Ca(OH)}_2$$

2   **0.16 M Mg(OH)$_2$.** This problem gives you Step 6 in the titration procedure and asks you to back-solve for the molarity of the base. Start by finding the number of moles of acid present in the solution by multiplying the molarity (0.50 M) by the volume (0.005 L), giving you 0.0025 mol. Next, examine the balanced neutralization reaction to determine the number of moles of magnesium hydroxide needed to neutralize 0.0025 mol HNO$_2$:

$$2HNO_2 + Mg(OH)_2 \rightarrow Mg(NO_2)_2 + 2H_2O$$

From the balanced equation, you can see that for every 2 mol of HNO$_2$ you need 1 mol of magnesium hydroxide to neutralize it:

$$\left(\frac{0.0025 \text{ mol HNO}_2}{1} \right) \left(\frac{1 \text{ mol Mg(OH)}_2}{2 \text{ mol HNO}_2} \right) = 1.25 \times 10^{-3} \text{ mol Mg(OH)}_2$$

Divide this value by the volume of base added (8.0 mL, or 0.008 L) to get the molarity:

$$\left(\frac{1.25 \times 10^{-3} \text{ mol Mg(OH)}_2}{0.008 \text{ L}} \right) = 0.16 \text{ M Mg(OH)}_2$$

3   Without an identity or a concentration, a chemist could still determine the number of moles per liter of the mystery acid or base. If he titrates an extra-mysterious acid with a base of known concentration until he achieves neutrality, then he knows the number of moles of acid in the solution (equal to the number of moles of base added). This information may even allow him to guess at its identity.

4   **pH of 4.7.** Ethanoic acid, also known as *acetic acid* or *vinegar,* would be an ideal buffer for a pH close to its pK_a value. You're given the K_a of ethanoic acid in the example problem (it's $K_a = 1.8 \times 10^{-5}$), so simply take the negative logarithm of that value to get your pK_a value and therefore your pH:

$$-\log \left(1.8 \times 10^{-5} \right) = pK_a = 4.7$$

5 You're given a pH and a pK_a, which suggests that you need to use the Henderson-Hasselbalch equation. You have the total concentration of acid and conjugate base, but you don't know either of the concentrations in the equation individually, so begin by solving for their ratio:

$$\frac{[A^-]}{[HA]} = 10^{(pH-pKa)} = 10^{(7-6.8)} = 1.6$$

So you have 1.6 mol of base for every 1 mol of acid. Expressing the amount of base as a fraction of the total amount of acid and base therefore gives you

$$\frac{[A^-]}{[HA]+[A^-]} = \frac{1.6}{1+1.6} = 0.62$$

Multiply this number by the molarity of the solution (0.2 M) to give you the molarity of the basic solution, or 0.12 M. The molarity of the acid is therefore 0.2 M − 0.12 M = 0.08 M.

This means that to prepare the buffer, you must take 0.2 mol of carbonic acid and add it to somewhat less than 1,000 mL of water (say, 800 mL). To this you must add enough of a strong base such as sodium hydroxide (NaOH) to force the proper proportion of the carbonic acid solution to dissociate into its conjugate base. Because you want to achieve a base concentration of 0.12 M, you should add 120 mL of NaOH. Finally, you should add enough water to achieve your final volume of 1,000 mL.

6 **For silver (I) chromate:** $K_{sp} = [Ag^+]^2 [CrO_4^{2-}]$; **for strontium sulfate:** $K_{sp} = [Sr^{2+}] [SO_4^{2-}]$. To determine the solubility product constants of these solutions, you first need to write an equation for their dissociation in water (see Chapter 8 for details):

$$Ag_2CrO_4 \rightarrow 2Ag^+(aq) + CrO_4^{2-}(aq)$$
$$SrSO_4 \rightarrow Sr^{2+}(aq) + SO_4^{2-}(aq)$$

Raising the concentration of the products to the power of their coefficients in these balanced reactions yields the K_{sp} for each:

$$K_{sp} = [Ag^+]^2 [CrO_4^{2-}]$$
$$K_{sp} = [Sr^{2+}] [SO_4^{2-}]$$

7 4×10^{-6}. You write an expression for the K_{sp} of silver (I) chromate in Problem 6. Solve the equation for the chromate concentration by dividing K_{sp} by the silver ion concentration:

$$[CrO_4^{2-}] = \frac{K_{sp}}{[Ag^+]^2} = \frac{(1.1 \times 10^{-12})}{(5 \times 10^{-4})^2} = 4 \times 10^{-6}$$

Chapter 18

Accounting for Electrons in Redox

. . . .

In This Chapter
▶ Keeping an eye on electrons by using oxidation numbers
▶ Balancing redox reactions in the presence of acid
▶ Balancing redox reactions in the presence of base

. . . .

*I*n chemistry, electrons get all the action. Among other things, electrons can transfer between reactants during a reaction. Reactions like these are called *oxidation-reduction* reactions, or *redox* reactions for short. Redox reactions are as important as they are common. But they're not always obvious. In this chapter, you find out how to recognize redox reactions and how to balance the equations that describe them.

Oxidation Numbers: Keeping Tabs on Electrons

If electrons move between reactants during redox reactions, then it should be easy to recognize those reactions just by noticing changes in charge, right? Sometimes. The following two reactions are both redox reactions:

$$Mg(s) + 2H^+(aq) \rightarrow Mg^{2+}(aq) + H_2(g)$$

$$2H_2(g) + O_2 \rightarrow 2H_2O(g)$$

In the first reaction, it's obvious that electrons are transferred from magnesium to hydrogen. There are only two reactants, and magnesium is neutral as a reactant but positively charged as a product. The charges can change in this way only through electron transfer. In the second reaction, electrons are transferred from hydrogen to oxygen, but the transfer isn't as obvious. Because you'll encounter both types of reactions, you need a simple way to keep tabs on electrons as they transfer between reactants. In short, you need oxidation numbers.

Oxidation numbers are tools to keep track of electrons. Sometimes an oxidation number describes the actual charge on an atom. Other times, an oxidation number describes an imaginary sort of charge — the charge an atom would have if all its bonding partners left town, taking their own electrons with them. In ionic compounds, oxidation numbers come closest to describing actual atomic charge. The description is less apt in covalent compounds, in which electrons are less clearly "owned" by one atom or another. (See Chapter 6 for the basics on different types of compounds.) The point is this: Oxidation numbers are useful tools, but they aren't direct descriptions of physical reality.

Here are some basic rules for figuring out an atom's oxidation number:

- An atom in elemental form (no charge, bonded to nothing or itself) has an oxidation number of 0. Therefore, the oxidation number of both $Mg(s)$ and $O_2(g)$ is 0.

- Single-atom *(monatomic)* ions have an oxidation number equal to their charge. So the oxidation number of $Mg^{2+}(aq)$ is +2, and the oxidation number of $Cl^-(aq)$ is –1.

- In a neutral compound, oxidation numbers add up to 0. In a charged compound, oxidation numbers add up to the compound's charge.

- In compounds, oxygen usually has an oxidation number of –2. An annoying exception is the peroxides, like H_2O_2, in which oxygen has an oxidation number of –1.

- In compounds, hydrogen has an oxidation number of +1 when it bonds to nonmetals (as in H_2O) and an oxidation number of –1 when it bonds to metals (as in NaH).

- In compounds, the following rules apply for various families:

  - Group IA atoms (alkali metals) have an oxidation number of +1.

  - Group IIA atoms (alkaline earth metals) have an oxidation number of +2.

  - In Group IIIA, Al and Ga atoms have an oxidation number of +3.

  - Group VIIA atoms (halogens) usually have an oxidation number of –1.

  Notice that the oxidation states for the families such as the alkali metals, alkaline earth metals, and so on correspond to the ionic charges those elements have.

You can deduce the oxidation numbers of other atoms in compounds (especially those of transition metals) by taking into account these basic rules as well as the compound's overall charge.

By applying these rules to chemical reactions, you can discern who supplies electrons to whom:

- The chemical species that loses electrons is *oxidized* and acts as the *reducing agent* (or *reductant*). (In short, oxidized = reducing agent.)

- The species that gains electrons is *reduced* and acts as the *oxidizing agent* (or *oxidant*). (In short, reduced = oxidizing agent.)

Chemistry terminology can be somewhat confusing (what an understatement!) as electrons are transferred back and forth and you try to keep track of everything with terms like *reduction, oxidizing,* and the like. The phrase OIL RIG is a helpful little tool to remember what means what:

- **OIL — Oxidation is loss:** Simply put, oxidation is a loss of electrons. The element being oxidized is giving away electrons.

- **RIG — Reduction is gain:** Reduction is a gain of electrons for an element. The element that accepts electrons in a reaction is said to be reduced.

All redox reactions have both an oxidizing agent and a reducing agent. Although a given beaker can contain many different oxidizing agent–reducing agent pairs, each pair constitutes its own redox reaction.

Oxidation always means an increase in oxidation number. *Reduction* always means a decrease in oxidation number.

0. Use oxidation numbers to identify the oxidizing and reducing agents in the following chemical reaction:

$$Cr_2O_3(s) + 2Al(s) \rightarrow 2Cr(s) + Al_2O_3(s)$$

A. **$Cr_2O_3(s)$ is the oxidizing agent, and Al(s) is the reducing agent.** The keys here are to recall that atoms in elemental form (like solid Al and solid Cr) have oxidation numbers of 0 and that oxygen typically has an oxidation number of –2 in compounds. The oxidation number breakdown (shown in the following figure) reveals that Al(s) is oxidized to $Al_2O_3(s)$ and $Cr_2O_3(s)$ is reduced to Cr(s). So Al(s) is the reducing agent, and $Cr_2O_3(s)$ is the oxidizing agent.

You can tell that Al(s) is oxidized because the oxidation number of Al is 0 in Al(s) but is +3 in $Al_2O_3(s)$. This corresponds to a loss of three electrons because the charge changes from 0 to +3. You can tell that $Cr_2O_3(s)$ is reduced because the oxidation number of Cr is +3 in $Cr_2O_3(s)$ but is 0 in Cr(s), which corresponds to a gain of three electrons to reduce the charge to zero.

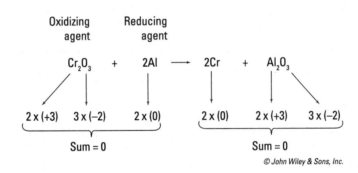

© John Wiley & Sons, Inc.

---

1. Use oxidation numbers to identify the oxidizing and reducing agents in the following chemical reaction:

$$MnO_2(s) + 2H^+(aq) + NO_2^-(aq) \rightarrow NO_3^-(aq) + Mn^{2+}(aq) + H_2O(l)$$

Solve It

---

2. Use oxidation numbers to identify the oxidizing and reducing agents in the following chemical reaction:

$$H_2S(aq) + 2NO_3(aq) + 2H^+(aq) \rightarrow S(s) + 2NO_2(g) + 2H_2O(l)$$

Solve It

Balancing Redox Reactions under Acidic Conditions

When you balance a chemical reaction equation, the primary concern is to obey the principle of conservation of mass: The total mass of the reactants must equal the total mass of the products. (See Chapter 8 if you need to review this process.) In redox reactions, you must obey a second principle as well: the *conservation of charge.* The total number of electrons lost must equal the total number of electrons gained. In other words, you can't just leave electrons lying around. The universe is finicky about that type of thing.

Sometimes simply balancing a redox reaction with an eye to mass results in a charge-balanced equation as well. Like a string of green traffic lights when you're driving, that's a lovely thing when it occurs, but you can't count on it. So it's best to have a go-to system for balancing redox reactions. The details of that system depend on whether the reaction occurs in acidic or basic conditions — in the presence of excess H^+ or excess OH^-. Both variations use *half-reactions,* incomplete parts of the total reaction that reflect either oxidation or reduction alone. (We introduce the basics of acids and bases in Chapter 16.)

Here's a summary of the method for balancing a redox reaction equation for a reaction under acidic conditions (excess H^+) (see the next section for details on balancing a reaction under basic conditions):

1. **Separate the reaction equation into the oxidation half-reaction and the reduction half-reaction. Use oxidation numbers to identify these component half-reactions.**

2. **Balance the half-reactions separately by adding coefficients, temporarily ignoring O and H atoms.**

3. **Turn your attention to the O and H atoms. Balance the half-reactions separately, using H_2O to add O atoms and using H^+ to add H atoms.**

4. **Balance the half-reactions separately for charge by adding electrons (e^-).**

5. **Balance the charge of the half-reactions with respect to each other by multiplying the reactions so that the total number of electrons is the same in each half-reaction.**

6. **Reunite the half-reactions into a complete redox reaction equation.**

7. **Simplify the equation by canceling items that appear on both sides of the arrow. Be sure your electrons cancel.**

Balancing a redox reaction can seem overwhelming, but it really isn't. Just follow the steps, practice a few problems, and you won't have any trouble. As a galactically renowned book once said, "Don't Panic."

0. Balance the following redox reaction equation, assuming acidic conditions:

$$NO_2^- + Br^- \rightarrow NO + Br_2$$

A. Simply go through the steps, 1–7.

1. Divide the equation into half-reactions for oxidation and reduction by using oxidation numbers.

$$Br^- \rightarrow Br_2 \quad \text{(oxidation)}$$
$$NO_2^- \rightarrow NO \quad \text{(reduction)}$$

2. Balance the half-reactions by adding coefficients, temporarily ignoring O and H.

$$2Br^- \rightarrow Br_2$$
$$NO_2^- \rightarrow NO$$

3. Balance the half-reactions for O and H by adding H_2O and H^+, respectively.

$$2Br^- \rightarrow Br_2$$
$$2H^+ + NO_2^- \rightarrow NO + H_2O$$

4. Balance the charge within each half-reaction by adding electrons (e^-).

$$2Br^- \rightarrow Br_2 + 2e^-$$
$$1e^- + 2H^+ + NO_2^- \rightarrow NO + H_2O$$

5. Balance the charge of the half-reactions with respect to each other (in other words, make the electrons equal to each other).

$$2Br^- \rightarrow Br_2 + 2e^-$$
$$2e^- + 4H^+ + 2NO_2^- \rightarrow 2NO + 2H_2O$$

6. Add the half-reactions, reuniting them within the total reaction equation.

$$2e^- + 2Br^- + 4H^+ + 2NO_2^- \rightarrow 2NO + 2H_2O + Br_2 + 2e^-$$

7. Simplify by canceling items that appear on both sides of the equation.

$$2Br^- + 4H^+ + 2NO_2^- \rightarrow 2NO + 2H_2O + Br_2$$

---

3. Balance the following redox reaction equation, assuming acidic conditions:

$$Fe^{3+} + SO_2 \rightarrow Fe^{2+} + SO_4^{2-}$$

Solve It

4. Balance the following redox reaction equation, assuming acidic conditions:

$$Ag^+ + Be \rightarrow Ag + Be_2O_3^{2-}$$

Solve It

5. Balance the following redox reaction equation, assuming acidic conditions:

$$Cr_2O_7{}^{2-} + Bi \rightarrow Cr^{3+} + BiO^+$$

Solve It

Balancing Redox Reactions under Basic Conditions

Yes, balancing redox equations can involve quite a lot of bookkeeping. Not much can be done to remedy that unfortunate fact. But here's the good news: The process for balancing redox equations under basic conditions is 90 percent identical to the one used for balancing under acidic conditions in the preceding section. In other words, master one, and you've mastered both.

Here's how easy it is to adapt your balancing method for basic conditions:

1–7. Perform Steps 1–7 as described in the preceding section for balancing under acidic conditions.

8. Observe where H+ is present in the resulting equation. Add an identical amount of OH– to both sides of the equation so that all the H+ is "neutralized," becoming water.

9. Cancel any amounts of H_2O that appear on both sides of the equation.

That's it. Really.

Q. Balance the following redox reaction equation, assuming basic conditions:

$$Cr^{2+} + Hg \rightarrow Cr + HgO$$

A. Begin balancing as if the reaction occurred under acidic conditions, and then neutralize any H+ ions by adding OH– equally to both sides. Finally, cancel any excess H_2O molecules.

1. **Divide the equation into half-reactions for oxidation and reduction by using oxidation numbers.**

$$Hg \rightarrow HgO \quad \text{(oxidation)}$$
$$Cr^{2+} \rightarrow Cr \quad \text{(reduction)}$$

2. **Balance the half-reactions, temporarily ignoring O and H. (They're already balanced here.)**

$$Hg \rightarrow HgO$$
$$Cr^{2+} \rightarrow Cr$$

3. **Balance the half-reactions for O and H by adding H_2O and H^+, respectively.**

$$Hg + H_2O \rightarrow HgO + 2H^+$$
$$Cr^{2+} \rightarrow Cr$$

4. **Balance the charge within each half-reaction by adding electrons (e⁻).**

$$Hg + H_2O \rightarrow HgO + 2H^+ + 2e^-$$
$$Cr^{2+} + 2e^- \rightarrow Cr$$

5. **Balance the charge of the half-reactions with respect to each other. (They're already balanced here.)**

$$Hg + H_2O \rightarrow HgO + 2H^+ + 2e^-$$
$$Cr^{2+} + 2e^- \rightarrow Cr$$

6. **Add the half-reactions, reuniting them within the total reaction equation.**

$$2e^- + Cr^{2+} + Hg + H_2O \rightarrow HgO + 2H^+ + Cr + 2e^-$$

7. **Simplify by canceling items that appear on both sides of the equation.**

$$Cr^{2+} + Hg + H_2O \rightarrow HgO + 2H^+ + Cr$$

8. **Neutralize H^+ by adding sufficient and equal amounts of OH⁻ to both sides.**

$$Cr^{2+} + Hg + H_2O + 2OH^- \rightarrow HgO + Cr + 2H_2O$$

9. **Simplify by canceling H_2O as possible from both sides.**

$$Cr^{2+} + Hg + 2OH^- \rightarrow HgO + Cr + H_2O$$

6. Balance the following redox reaction equation, assuming basic conditions:

$$Cl_2 + Sb \rightarrow Cl^- + SbO^+$$

Solve It

7. Balance the following redox reaction equation, assuming basic conditions:

$$SO_4^{2-} + ClO_2^- \rightarrow SO_2 + ClO_2$$

Solve It

8. Balance the following redox reaction equation, assuming basic conditions:

$$BrO_3^- + PbO \rightarrow Br_2 + PbO_2$$

Solve It

Answers to Questions on Electrons in Redox

Among chemistry students, redox reactions have a bad reputation because of perceptions that they're difficult to understand and even more difficult to balance. But the perception is only a perception. The whole process boils down to the following principles:

- ✔ Remember that balancing redox reaction equations is exactly like balancing other equations; you simply have one more component to balance: the electron.
- ✔ Use oxidation numbers to discern the oxidation and reduction half-reactions.
- ✔ Under acidic conditions, balance O and H atoms by adding H_2O and H^+.
- ✔ Under basic conditions, balance as you do with acidic conditions but then neutralize any H^+ by adding OH^-.

So relax. Well, relax after you check your work.

1 **The oxidizing agent is MnO_2, and the reducing agent is NO_2^-.** The oxidation number of Mn is +4 in the MnO_2 reactant and is +2 in the Mn^{2+} product. The oxidation number of N is +3 in the NO_2^- reactant and is +5 in the NO_3^- product.

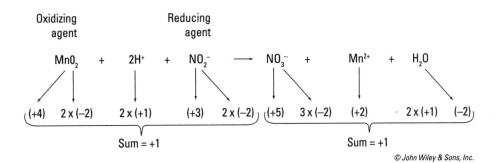

© John Wiley & Sons, Inc.

2 **The oxidizing agent is NO_3^-, and the reducing agent is H_2S.** The oxidation number of S is –2 in the H_2S reactant and is 0 in the S product. The oxidation number of N is +5 in the NO_3^- reactant and is +4 in the NO_2 product.

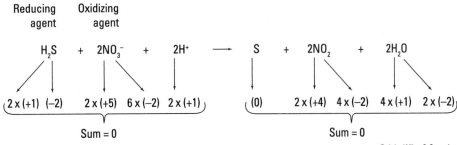

© John Wiley & Sons, Inc.

3   Follow the seven steps for balancing a redox reaction equation under acidic conditions:

1. **Divide the equation into half-reactions for oxidation and reduction by using oxidation numbers.**

$$SO_2 \rightarrow SO_4^{2-} \quad \text{(oxidation)}$$
$$Fe^{3+} \rightarrow Fe^{2+} \quad \text{(reduction)}$$

2. **Balance the half-reactions by adding coefficients, temporarily ignoring O and H.**

(already complete)

3. **Balance the half-reactions for O and H by adding H_2O and H^+, respectively.**

$$2H_2O + SO_2 \rightarrow SO_4^{2-} + 4H^+$$
$$Fe^{3+} \rightarrow Fe^{2+}$$

4. **Balance the charge within each half-reaction by adding electrons (e^-).**

$$2H_2O + SO_2 \rightarrow SO_4^{2-} + 4H^+ + 2e^-$$
$$1e^- + Fe^{3+} \rightarrow Fe^{2+}$$

5. **Balance the charge of the half-reactions with respect to each other (in other words, make the electrons equal to each other).**

$$2H_2O + SO_2 \rightarrow SO_4^{2-} + 4H^+ + 2e^-$$
$$2e^- + 2Fe^{3+} \rightarrow 2F^{2+}$$

6. **Add the half-reactions, reuniting them within the total reaction equation.**

$$2e^- + 2H_2O + 2Fe^{3+} + SO_2 \rightarrow 2Fe^{2+} + SO_4^{2-} + 4H^+ + 2e^-$$

7. **Simplify by canceling items that appear on both sides of the equation.**

$$2H_2O + 2Fe^{3+} + SO_2 \rightarrow 2Fe^{2+} + SO_4^{2-} + 4H^+$$

4   Follow the seven steps for balancing a redox reaction equation under acidic conditions:

1. $Be \rightarrow Be_2O_3^{2-}$ (oxidation)
   $Ag^+ \rightarrow Ag$ (reduction)

2. $2Be \rightarrow Be_2O_3^{2-}$
   $Ag^+ \rightarrow Ag$

3. $3H_2O + 2Be \rightarrow Be_2O_3^{2-} + 6H^+$
   $Ag^+ \rightarrow Ag$

4. $3H_2O + 2Be \rightarrow Be_2O_3^{2-} + 6H^+ + 4e^-$
   $1e^- + Ag^+ \rightarrow Ag$

5. $3H_2O + 2Be \rightarrow Be_2O_3^{2-} + 6H^+ + 4e^-$
   $4e^- + 4Ag^+ \rightarrow 4Ag$

6. $4e^- + 3H_2O + 4Ag^+ + 2Be \rightarrow 4Ag + Be_2O_3^{2-} + 6H^+ + 4e^-$

7. $3H_2O + 4Ag^+ + 2Be \rightarrow 4Ag + Be_2O_3^{2-} + 6H^+$

5 Follow the seven steps for balancing a redox reaction equation under acidic conditions:

1. $Bi \rightarrow BiO^+$ (oxidation)
   $Cr_2O_7^{2-} \rightarrow Cr^{3+}$ (reduction)

2. $Bi \rightarrow BiO^+$
   $Cr_2O_7^{2-} \rightarrow 2Cr^{3+}$

3. $H_2O + Bi \rightarrow BiO^+ + 2H^+$
   $14H^+ + Cr_2O_7^{2-} \rightarrow 2Cr^{3+} + 7H_2O$

4. $H_2O + Bi \rightarrow BiO^+ + 2H^+ + 3e^-$
   $6e^- + 14H^+ + Cr_2O_7^{2-} \rightarrow 2Cr^{3+} + 7H_2O$

5. $2H_2O + 2Bi \rightarrow 2BiO^+ + 4H^+ + 6e^-$
   $6e^- + 14H^+ + Cr_2O_7^{2-} \rightarrow 2Cr^{3+} + 7H_2O$

6. $6e^- + 14H^+ + Cr_2O_7^{2-} + 2H_2O + 2Bi \rightarrow 2Cr^{3+} + 7H_2O + 2BiO^+ + 4H^+ + 6e^-$

7. $10H^+ + Cr_2O_7^{2-} + 2Bi \rightarrow 2Cr^{3+} + 5H_2O + 2BiO^+$

6 Follow the nine steps for balancing a redox reaction equation under basic conditions:

1. **Divide the equation into half-reactions for oxidation and reduction by using oxidation numbers.**

$$Sb \rightarrow SbO^+ \quad \text{(oxidation)}$$
$$Cl_2 \rightarrow Cl^- \quad \text{(reduction)}$$

2. **Balance the half-reactions, temporarily ignoring O and H.**

$$Sb \rightarrow SbO^+$$
$$Cl_2 \rightarrow 2Cl^-$$

3. **Balance the half-reactions for O and H by adding H_2O and H^+, respectively.**

$$Sb + H_2O \rightarrow SbO^+ + 2H^+$$
$$Cl_2 \rightarrow 2Cl^-$$

4. **Balance the charge within each half-reaction by adding electrons (e⁻).**

$$Sb + H_2O \rightarrow SbO^+ + 2H^+ + 3e^-$$
$$2e^- + Cl_2 \rightarrow 2Cl^-$$

5. **Balance the charge of the half-reactions with respect to each other.**

$$2Sb + 2H_2O \rightarrow 2SbO^+ + 4H^+ + 6e^-$$
$$6e^- + 3Cl_2 \rightarrow 6Cl^-$$

6. **Add the half-reactions, reuniting them within the total reaction equation.**

$$6e^- + 3Cl_2 + 2Sb + 2H_2O \rightarrow 6Cl^- + 2SbO^+ + 4H^+ + 6e^-$$

7. **Simplify by canceling items that appear on both sides of the equation.**

$$3Cl_2 + 2Sb + 2H_2O \rightarrow 6Cl^- + 2SbO^+ + 4H^+$$

8. Neutralize H⁺ by adding sufficient and equal amounts of OH⁻ to both sides.

$$3Cl_2 + 2Sb + 2H_2O + 4OH^- \rightarrow 6Cl^- + 2SbO^+ + 4H_2O$$

9. Simplify by canceling H₂O as possible from both sides.

$$3Cl_2 + 2Sb + 4OH^- \rightarrow 6Cl^- + 2SbO^+ + 2H_2O$$

7 Follow the nine steps for balancing a redox reaction equation under basic conditions:

1. $ClO_2^- \rightarrow ClO_2$(oxidation)

    $SO_4^{2-} \rightarrow SO_2$(reduction)

2. (Already complete.)

3. $ClO_2^- \rightarrow ClO_2$

    $SO_4^{2-} + 4H^+ \rightarrow SO_2 + 2H_2O$

4. $ClO_2^- \rightarrow ClO_2 + 1e^-$

    $2e^- + SO_4^{2-} + 4H^+ \rightarrow SO_2 + 2H_2O$

5. $2ClO_2^- \rightarrow 2ClO_2 + 2e^-$

    $2e^- + SO_4^{2-} + 4H^+ \rightarrow SO_2 + 2H_2O$

6. $2e^- + SO_4^{2-} + 4H^+ + 2ClO_2^- \rightarrow SO_2 + 2H_2O + 2ClO_2 + 2e^-$

7. $SO_4^{2-} + 4H^+ + 2ClO_2^- \rightarrow SO_2 + 2H_2O + 2ClO_2$

8. $SO_4^{2-} + 4H_2O + 2ClO_2^- \rightarrow SO_2 + 2H_2O + 2ClO_2 + 4OH^-$

9. $SO_4^{2-} + 2H_2O + 2ClO_2^- \rightarrow SO_2 + 2ClO_2 + 4OH^-$

8 Follow the nine steps for balancing a redox reaction equation under basic conditions:

1. $PbO \rightarrow PbO_2$(oxidation)

    $BrO_3^- \rightarrow Br_2$(reduction)

2. $PbO \rightarrow PbO_2$

    $2BrO_3^- \rightarrow Br_2$

3. $H_2O + PbO \rightarrow PbO_2 + 2H^+$

    $12H^+ + 2BrO_3^- \rightarrow Br_2 + 6H_2O$

4. $H_2O + PbO \rightarrow PbO_2 + 2H^+ + 2e^-$

    $10e^- + 12H^+ + 2BrO_3^- \rightarrow Br_2 + 6H_2O$

5. $5H_2O + 5PbO \rightarrow 5PbO_2 + 10H^+ + 10e^-$

    $10e^- + 12H^+ + 2BrO_3^- \rightarrow Br_2 + 6H_2O$

6. $10e^- + 12H^+ + 2BrO_3^- + 5H_2O + 5PbO \rightarrow Br_2 + 6H_2O + 5PbO_2 + 10H^+ + 10e^-$

7. $2H^+ + 2BrO_3^- + 5PbO \rightarrow Br_2 + H_2O + 5PbO_2$

8. $2H_2O + 2BrO_3^- + 5PbO \rightarrow Br_2 + H_2O + 5PbO_2 + 2OH^-$

9. $H_2O + 2BrO_3^- + 5PbO \rightarrow Br_2 + 5PbO_2 + 2OH^-$

Chapter 19

Galvanizing Yourself to Do Electrochemistry

. . . .

In This Chapter

▶ Understanding voltaic cells, anodes, and cathodes

▶ Figuring standard reduction potentials and electromotive force

▶ Zapping current into electrolytic cells

. . . .

*A*lthough we're sure that you're thrilled to hear that redox reactions are the driving force behind the creation of rust and the greening of the copper domes on cathedrals, you should know that you don't need to sit around watching metal rust to see redox reactions in action. In fact, they play a very important role in your everyday life. The redox reactions that you find out about in Chapter 18 are the silent partner backing the unstoppable Energizer Bunny. That's right! Redox reactions are the essential chemistry behind the inner workings of the all-important battery. In fact, an entire branch of chemistry, called *electrochemistry,* centers on the study of electrochemical cells, which create electrical energy from chemical energy. We introduce the basics in this chapter.

Identifying Anodes and Cathodes

The energy provided by batteries is created in a unit called a *voltaic* or *galvanic cell.* Many batteries use a number of voltaic cells wired in series, and others use a single cell. Voltaic cells harness the energy released in a redox reaction and transform it into electrical energy. A voltaic cell is created by connecting two metals called *electrodes* in solution with an external circuit. In this way, the reactants aren't in direct contact, but they can transfer electrons to one another through an external pathway, fueling the redox reaction.

The electrode that undergoes oxidation is called the *anode.* You can easily remember this if you burn the phrase "an ox" (for *an*ode *ox*idation) into your memory. The phrase "red cat" is equally useful for remembering what happens at the other electrode, called the *cathode,* where the reduction reaction occurs.

Electrons created in the oxidation reaction at the anode of a voltaic cell flow along an external circuit to the cathode, where they fuel the reduction reaction taking place there. We use the spontaneous reaction between zinc and copper as an example of a voltaic cell here, but you should realize that many powerful redox reactions power many types of batteries, so they're not limited to reactions between copper and zinc.

Zinc metal reacts spontaneously with an aqueous solution of copper sulfate when they're placed in direct contact. Zinc, being a more reactive metal than copper (it's higher on the activity series of metals presented in Chapter 8), displaces the copper ions in solution. The displaced copper deposits itself as pure copper metal on the surface of the dissolving zinc strip. At first, the reaction may appear to be a simple single replacement reaction, but it's also a redox reaction.

The oxidation of zinc proceeds as $Zn(s) \rightarrow Zn^{2+}(aq) + 2e^-$. The two electrons created in this oxidation of zinc are consumed by the copper in the reduction half of the reaction $Cu^{2+}(aq) + 2e^- \rightarrow Cu(s)$. This makes the total reaction $Zn(s) + Cu^{2+}(aq) + 2e^- \rightarrow Zn^{2+}(aq) + 2e^- + Cu(s)$. The electron duo appears on both sides of this combined reaction, so the reaction is really just $Zn(s) + Cu^{2+}(aq) \rightarrow Zn^{2+}(aq) + Cu(s)$.

This reaction takes place when zinc and copper are in direct contact, but as we explain earlier in this section, a voltaic cell is created by connecting the two reactants by an external pathway. Only the electrons created at the anode in the oxidation reaction can travel to the reduction half of the reaction along this external pathway. A voltaic cell using this same oxidation-reduction reaction between copper and zinc is shown in Figure 19-1, which we examine piece by piece.

- ✔ **Anode:** Zinc (Zn) is being oxidized at the anode, which is labeled with a negative sign. This doesn't mean that the anode is negatively charged. Rather, it means that electrons are being created there. The oxidation of zinc releases Zn^{2+} cations into the solution as well as two electrons that flow along the circuit to the cathode. This oxidation thus results in an increase of Zn^{2+} ions in the solution and a decrease in the mass of the zinc metal anode.

- ✔ **Cathode:** The electrons created by the oxidation of zinc at the anode fuel the reduction of copper (Cu) at the cathode. This pulls Cu^{2+} from solution and deposits more Cu metal on the cathode. The result is the exact opposite of the effect occurring at the anode: The solution becomes less concentrated as Cu^{2+} ions are used up in the reduction reaction, and the electrode gains mass as Cu metal is deposited.

- ✔ **Salt bridge:** Figure 19-1 also contains a U-shaped tube connecting the two solutions. This is called the *salt bridge,* and it serves to correct the charge imbalance created as the anode releases more and more cations into its solution, resulting in a net positive charge, and as the cathode uses up more and more of the cations in its solution, leaving it with a net negative charge.

    The salt bridge contains an electrolytic salt (in this case, $NaNO_3$). A good salt bridge is created with an electrolyte whose component ions won't react with the ions already in solution. The salt bridge functions by sucking up the extra NO_3^- ions in the cathode solution and depositing NO_3^- ions from the other end of the bridge into the anode solution. It also sucks up the excess positive charge at the anode by absorbing Zn^{2+} ions and depositing its own cation, Na^+, into the cation solution. This is necessary because the solutions must have balanced charges for the redox reaction to continue. The salt bridge also completes the cell's circuit by allowing for the flow of charge back to the anode.

    A useful way to remember how ions flow in the salt bridge is that anions travel to the anode, while cations travel to the cathode.

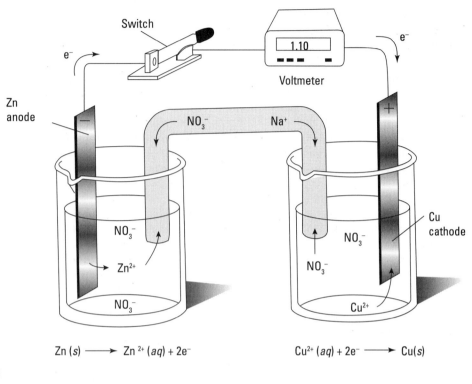

Zn anode

Switch

e⁻

1.10

Voltmeter

e⁻

NO_3^- Na^+

NO_3^-

Zn^{2+}

NO_3^-

NO_3^-

NO_3^-

Cu^{2+}

Cu cathode

$Zn\,(s) \longrightarrow Zn^{2+}\,(aq) + 2e^-$ $Cu^{2+}\,(aq) + 2e^- \longrightarrow Cu(s)$

Figure 19-1:
A voltaic
cell.

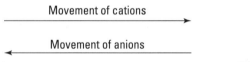

Movement of cations

Movement of anions

© John Wiley & Sons, Inc.

These voltaic cells can't run forever, however. The loss of mass at the zinc anode will eventually exhaust the supply of zinc, and the redox reaction won't be able to continue. This phenomenon is why most batteries run out over time. Rechargeable batteries take advantage of a reverse reaction to resupply the anode, but many redox reactions don't allow for this, so rechargeable batteries must be made of very specific reactants.

The voltage provided by a battery depends largely on the materials that make up the two electrodes. The reactions carried out in the 1.5 V batteries that power your calculator, flashlight, or MP3 player, for example, are often reactions between carbon and zinc chloride or between zinc and manganese dioxide. The high-voltage, long-lived batteries that power your computer, pacemaker, or watch, on the other hand, generally have anodes composed of a lithium compound, which can provide roughly twice the voltage of many other batteries and are longer-lived (certainly a desirable quality in the battery that powers a pacemaker). Although many batteries contain a single voltaic cell, a number of battery types that require high voltages, such as car batteries, harness the power of multiple voltaic cells by wiring them in series with one another. A 12 V car battery, for example, is created by wiring six 2 V voltaic cells together.

Voltage is measured by attaching a voltmeter to a circuit, as in Figure 19-1. When the switch is closed, the voltmeter reads the potential difference between the anode and cathode of the cell.

Q. A voltaic cell harnesses the reaction $2Al(s) + 3Sn^{2+}(aq) \rightarrow 2Al^{3+}(aq) + 3Sn(s)$. Which metal makes up the anode, and which makes up the cathode?

A. **The aluminum makes up the anode, and the tin makes up the cathode.** The best way to start is to add the spectator electrons to both sides of the equation to balance the positive charges of the cations. This gives you the equation

$$2Al(s) + 3Sn^{2+}(aq) + 6e^- \rightarrow 2Al^{3+}(aq) + 3Sn(s) + 6e^-$$

Next, isolate the oxidation and reduction half-reactions (as we explain in Chapter 18):

Oxidation half-reaction: $2Al(s) + 6e^- \rightarrow 2Al^{3+}(aq) + 6e^-$

Reduction half-reaction: $3Sn^{2+}(aq) + 6e^- \rightarrow 3Sn(s) + 6e^-$

Using your "an ox" and "red cat" mnemonics, you know that the anode is the site of aluminum (Al) oxidation, while the cathode is the site of tin (Sn) reduction. This idea is also apparent from the activity series of metals (see Chapter 8), which shows that aluminum is far more reactive than tin.

---

1. Write the oxidation and the reduction halves of a reaction between cadmium (II) and tin (II). Which makes up the anode, and which makes up the cathode? How do you know?

Solve It

2. Sketch a diagram of a cell composed of a bar of chromium in a chromium (III) nitrate solution and a bar of silver in a silver (II) nitrate solution with a salt bridge composed of potassium nitrate. Trace the movement of all ions and label the anode and the cathode. Make your diagram similar to Figure 19-1.

Solve It

Calculating Electromotive Force and Standard Reduction Potentials

In the preceding section, we take for granted that charge can flow through the external circuit connecting the two halves of a voltaic cell, but what actually causes this flow of charge? The answer lies in a concept called *potential energy.* When a difference in potential energy is established between two locations, an object has a natural tendency to move from an area of higher potential energy to an area of lower potential energy.

When Wile E. Coyote places a large, round boulder at the top of a hill overlooking a road on which he knows his nemesis the Road Runner will soon be traveling, he takes it for granted that when he releases the boulder, it will simply roll down the hill. This is due to the difference in potential energy between the top of the hill, which has high potential energy, and the bottom, which has low potential energy (although Coyote will need to be far better-versed in physics if he's to time the crushing of his adversary properly).

In a similar though less diabolical manner, the electrons produced at the anode of a voltaic cell have a natural tendency to flow along the circuit to a location with lower potential: the cathode. This potential difference between the two electrodes causes the *electromotive force,* or *EMF,* of the cell. EMF is also often referred to as the *cell potential* and is denoted E_{cell}. The cell potential varies with temperature and concentration of products and reactants and is measured in *volts* (V). The *standard cell potential,* or $E°_{cell}$, is the E_{cell} that occurs when concentrations of solutions are all at 1 M and the cell is at standard temperature and pressure (STP).

Much like assigning enthalpies to pieces of a reaction and summing them using Hess's law (see Chapter 15), cell potentials can be tabulated by taking the difference between the standard potentials of the oxidation and reduction half-reactions separately. To do this properly, chemists had to choose either the oxidation or reduction half-reaction to be positive. They happened to assign positive potentials to reduction half-reactions and negative potentials to oxidation half-reactions. The standard potential at an electrode is therefore a measurement of its propensity to undergo a reduction reaction. As such, these potentials are often referred to as *standard reduction potentials.* They're calculated using the formula

$$E°_{cell} = E°_{red}(\text{cathode}) - E°_{ox}(\text{anode})$$

Table 19-1 lists some standard reduction potentials along with the reduction half-reactions associated with them. The table is ordered from the most negative $E°_{red}$ (most likely to oxidize) to the most positive $E°_{red}$ (most likely to be reduced). The reactions with negative $E°_{red}$ are therefore reactions that happen at the anode of a voltaic cell, while those with a positive $E°_{red}$ occur at the cathode.

Table 19-1	Reduction Half-Reactions and Standard Reduction Potentials
Reduction Half-Reaction	$E°_{red}$ **(in Volts)**
$Li^+(aq) + e^- \rightarrow Li(s)$	−3.04
$K^+(aq) + e^- \rightarrow K(s)$	−2.92
$Ca^{2+}(aq) + 2e^- \rightarrow Ca(s)$	−2.76
$Na^+(aq) + e^- \rightarrow Na(s)$	−2.71
$Zn^{2+}(aq) + 2e^- \rightarrow Zn(s)$	−0.76
$Fe^{2+}(aq) + 2e^- \rightarrow Fe(s)$	−0.44
$2H^+(aq) + 2e^- \rightarrow H_2(g)$	0
$Cu^{2+}(aq) + 2e^- \rightarrow Cu(s)$	+0.34
$I_2(s) + 2e^- \rightarrow 2I^-(aq)$	+0.54
$Ag^+(aq) + e^- \rightarrow Ag(s)$	+0.80
$Br_2(l) + 2e^- \rightarrow 2Br^-(aq)$	+1.06
$O_3(g) + 2H^+(aq) + 2e^- \rightarrow O_2(g) + H_2O(l)$	+2.07
$F_2(g) + 2e^- \rightarrow 2F^-(aq)$	+2.87

Note that not all the reactions in Table 19-1 show the reduction of solid metals, as in our examples so far. We've thrown in liquids and gases as well. Not every voltaic cell is fueled by a reaction taking place between the metals of the electrodes. Although the cathode itself must be made of a metal to allow for the flow of electrons, those electrons can be passed into a gas or a liquid to complete the reduction half-reaction. Examine Figure 19-2 for an example of such a cell, which includes a gaseous electrode.

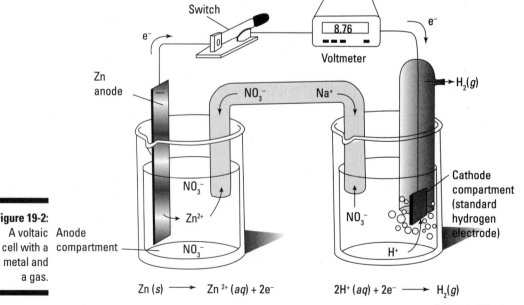

Figure 19-2:
A voltaic cell with a metal and a gas.

© John Wiley & Sons, Inc.

An equation relates standard cell potential to the electromotive force (EMF) of the cell. This equation, called the *Nernst equation,* is expressed as follows:

$$E_{cell} = E^\circ_{cell} - \left(\frac{RT}{nF}\right) \ln Q$$

Where Q is the reaction quotient (discussed in Chapter 14), n is the number of electrons transferred in the redox reaction, R is the universal gas constant 8.31 J/(mol·K), T is the temperature in kelvins, and F is the Faraday constant 9.65×10^4 coulombs/mol, where *coulomb* is a unit of electric charge. With this information, you can assign quantitative values to the EMFs of batteries. The equation also reveals that the EMF of a battery depends on temperature, which is why batteries are less likely to function well in the cold.

Q. What is the EMF of the cell shown in Figure 19-1 when it's at a temperature of 25°C if Cu^{2+} is 0.1 M and Zn^{2+} is 3.0 M?

A. **1.06 V.** To find the EMF, or E_{cell}, you first need to determine the standard cell potential. Do this by looking up the oxidation and reduction half-reactions in Table 19-1. The oxidation of zinc has an E°_{cell} of –0.76, and the reduction of copper has an E°_{cell} of 0.34. Recognizing that copper is the cathode and zinc is the anode, you can plug these values into the equation $E^\circ_{cell} = E^\circ_{red}(\text{cathode}) - E^\circ_{ox}(\text{anode})$ to get $E^\circ_{cell} = 0.34\ V + 0.76\ V = 1.10\ V$.

You then use the Nernst equation to calculate the EMF, so you have to determine the values of Q and n. You can find n by examining the oxidation and reduction half-reactions presented earlier in this chapter, which indicate that two electrons are exchanged in the process. Q is expressed as follows (if you can't recall how to calculate reaction quotients, see Chapter 14):

$$Q = \frac{[Zn^{2+}]}{[Cu^{2+}]} = \frac{[3.0]}{[0.1]} = 30$$

Next, write out the Nernst equation for this specific cell:

$$E_{cell} = E^\circ_{cell} - \left(\frac{\left(8.21\frac{J}{mol \cdot K}\right)(T)}{(n)\left(9.65 \times 10^4 \frac{coulombs}{mol}\right)}\right) \cdot \ln \frac{[Zn^{2+}]}{[Cu^{2+}]}$$

Plug in the known values for E°_{cell}, n, Cu^{2+}, and Zn^{2+}, convert the temperature to kelvins (just add 25 and 273), and solve:

$$E_{cell} = (1.10\ V) - \left(\frac{\left(8.21\frac{J}{mol \cdot K}\right)(298\ K)}{(2)\left(9.65 \times 10^4 \frac{coulombs}{mol}\right)}\right) \cdot \ln(30) = 1.06\ V$$

3. Using the two half-reactions fueling the cell in Figure 19-2, calculate the EMF of that cell at a temperature of 25°C and at 0°C, given concentrations of 10 M for H^+ and 0.01 M for Zn^{2+}.

Solve It

4. Is EMF constant over the life of a voltaic cell? Why or why not?

Solve It

Coupling Current to Chemistry: Electrolytic Cells

A positive standard cell potential tells you that the cathode is at a higher potential than the anode, and the reaction is therefore spontaneous. What do you do with a cell that has a negative $E°_{cell}$? Electrochemical cells that rely on such nonspontaneous reactions are called *electrolytic cells.* The redox reactions in electrolytic cells rely on a process called *electrolysis.* These reactions require that a current be passed through the solution, forcing it to split into components that then fuel the redox reaction. Such cells are created by applying a current source, such as a battery, to electrodes placed in a solution of *molten salt,* or salt heated until it melts. This splits the ions that make up the salt.

Cations have a natural tendency to migrate toward the negative anode, where they're oxidized, and anions migrate toward the positive cathode and are reduced. Thus, the "an ox" (anode oxidation) and "red cat" (reduction cathode) mnemonics are still valid.

Figure 19-3 shows an electrolytic cell using molten sodium chloride. A redox reaction between sodium and chlorine won't happen spontaneously, but the electrical energy produced by the battery provides the additional energy needed to fuel the reaction. In the process, chlorine anions are oxidized at the anode, creating chlorine gas, and sodium is reduced at the cathode and is deposited onto it as sodium metal.

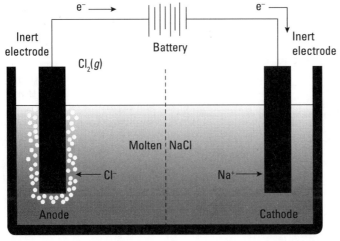

Figure 19-3:
An example of an electrolytic cell.

© John Wiley & Sons, Inc.

You can quantitatively analyze the amount of pure metal created in an electrolytic cell. First, you need to calculate the number of electrons that must be created to fuel electrolysis. To do this, write out the reduction half-reaction and determine how many electrons are needed to accomplish it. In the cell in Figure 19-3, reduction of sodium at the cathode occurs according to the equation $Na^+(aq) + e^- \rightarrow Na(s)$; thus, 1 mol of electrons can reduce aqueous sodium to produce 1 mol of sodium metal. If the cathode were reducing copper metal according to the reaction $Cu^{2+}(aq) + 2e^- \rightarrow Cu(s)$, it would require 2 mol of electrons for every mole of copper created, and so on.

To determine how many moles of metal have been deposited, you need to determine how many electrons have flowed through the circuit to fuel this reaction. With the following relationship, you can determine the amount of charge passing through the circuit (measured in coulombs, C, the standard unit of charge) by using the current provided by the power source and the amount of time the cell operates:

$$\text{Charge} = (\text{Current in amperes})(\text{Time in seconds})$$

This works out because the *ampere* (the standard unit of current, abbreviated A) is defined as 1 coulomb per second. Because this equation gives you the amount of charge that has passed through the circuit during its operating time, all that remains is to calculate the number of moles of electrons that make up that amount of charge. For this, you use the conversion factor 1 mol e^- = 96,500 C.

EXAMPLE

Q. How many grams of sodium metal would be deposited on the cathode of the electrolytic cell shown in Figure 19-3 if it were connected to a battery providing a 15 A current for 1.5 hours?

A. **19.32g Na(s).** Start by determining the amount of charge created in the cell. To do this, you must multiply the current by the time in seconds, which requires the conversion

$$\left(\frac{1.5 \text{ hr}}{1}\right)\left(\frac{60 \text{ min}}{1 \text{ hr}}\right)\left(\frac{60 \text{ sec}}{1 \text{ min}}\right)$$
$$= 5.4 \times 10^3 \text{ sec}$$

Plug this value for time into the equation Charge = (Current in amperes)(Time in seconds):

$$\text{Charge} = (15 \text{ A})\left(5.4 \times 10^3 \text{ sec}\right) = 8.1 \times 10^4 \text{ C}$$

Convert this charge to moles of electrons:

$$\left(\frac{8.1 \times 10^4 \text{ C}}{1}\right)\left(\frac{1 \text{ mol e}^-}{96,500 \text{ C}}\right) = 0.84 \text{ mol e}^-$$

Because you know that 1 mol of electrons can create 1 mol of sodium metal, this translates to 0.84 mol Na(s) produced. All that remains is to convert this value to grams using the gram atomic mass (see Chapter 7 for details):

$$\left(\frac{0.84 \text{ mol Na(s)}}{1}\right)\left(\frac{23.0 \text{ g Na}}{1 \text{ mol Na}}\right)$$
$$= 19.32 \text{ g Na(s)}$$

---

5. How much sodium would be deposited in the same amount of time as in the example question if the battery could supply a 50 A current?

Solve It

6. If an electrolytic cell containing molten aluminum (III) chloride were attached to a battery providing 8.0 A for 45 minutes, how many grams of pure aluminum metal would be deposited on the cathode in that time?

Solve It

Answers to Questions on Electrochemistry

Are you charged up to check your answers? Take a gander at this section to see how you did on the practice problems in this chapter.

1 You actually need to evaluate the second portion of this question first. A simple glance at the activity series of metals in Chapter 8 tells you that cadmium is more active than tin, so it must be cadmium that is oxidized and makes up the anode, and tin must be reduced and make up the cathode. The half-reactions are therefore

$$\text{Oxidation half-reaction: } Cd(s) \rightarrow Cd^{2+}(aq) + 2e^-$$

$$\text{Reduction half-reaction: } Sn^{2+}(aq) + 2e^- \rightarrow Sn(s)$$

2 Chromium is more active than silver (as we explain in Chapter 8), so chromium is oxidized in this reaction and silver is reduced according to the following half-reactions:

$$\text{Oxidation half-reaction: } Cr(s) \rightarrow Cr^{3+}(aq) + 3e^-$$

$$\text{Reduction half-reaction: } Ag^{2+}(aq) + 2e^- \rightarrow Ag(s)$$

The balanced redox reaction is therefore as follows (see Chapter 18 for details on balancing redox reactions):

$$2Cr(s) + 3Ag^{2+}(aq) + 6e^- \rightarrow 2Cr^{3+}(aq) + 6e^- + 3Ag(s)$$

Your diagram should therefore look like the following figure:

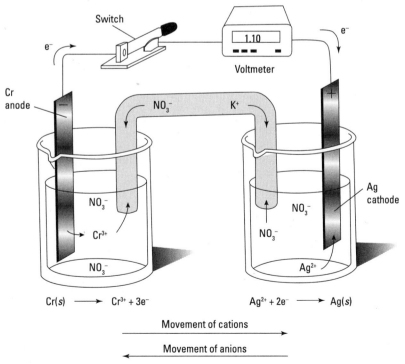

© John Wiley & Sons, Inc.

3 **0.848 V for 25°C and 0.840 V for 0°C.**

$$\text{Oxidation half-reaction: } Zn(s) \rightarrow Zn^{2+}(aq) + 2e^-$$

$$\text{Reduction half-reaction: } 2H^+(aq) + 2e^- \rightarrow H_2(g)$$

Table 19-1 shows that the potential for the oxidation half-reaction is –0.76 V and that the reduction half is 0 V. Plugging these values into the equation $E°_{cell} = E°_{red}(\text{cathode}) - E°_{ox}(\text{anode})$ gives you

$$E°_{cell} = 0 \text{ V} + 0.76 \text{ V} = 0.76 \text{ V}$$

The problem asks you to calculate the E_{cell} at $T = 298$ K (which is 25°C) and at $T = 273$ K (which is 0°C) and tells you that H+ = 10 M and Zn2+ = 0.01 M. The reaction quotient Q is

$$Q = \frac{[Zn^{2+}]}{[H^+]} = \frac{[0.01]}{[10]} = 0.001$$

Plugging all these values into the Nernst equation yields the following:

$$\text{For 25°C: } E_{cell} = 0.76 \text{ V} - \left(\frac{\left(8.21 \frac{J}{mol \cdot K} \right)(298 \text{ K})}{(2) \left(9.65 \times 10^4 \frac{coulombs}{mol} \right)} \right) \cdot \ln(0.001) = 0.848 \text{ V}$$

$$\text{For 0°C: } E_{cell} = 0.76 \text{ V} - \left(\frac{\left(8.21 \frac{J}{mol \cdot K} \right)(273 \text{ K})}{(2) \left(9.65 \times 10^4 \frac{coulombs}{mol} \right)} \right) \cdot \ln(0.001) = 0.840 \text{ V}$$

4 EMF isn't constant over the life of a voltaic cell, because the concentrations of the aqueous solutions are in flux. The anode solution's concentration increases over time, and the cathode solution's concentration decreases, changing the value of the reaction quotient and therefore the EMF.

5 **64.4 g Na(s).** To do this calculation, you only need to change the value of the current in the sample problem and follow the calculation through the rest of the way. Plugging the value of 50 A in for the current and keeping the same value for time in the equation Charge = (Current)(Time), you get

$$(50 \text{ A})(5.4 \times 10^3 \text{ s}) = 2.7 \times 10^5 \text{ C}$$

Convert this charge to moles of electrons:

$$\left(\frac{2.7 \times 10^5 \text{ C}}{1} \right) \left(\frac{1 \text{ mol } e^-}{96,500 \text{ C}} \right) = 2.80 \text{ mol } e^-$$

Convert 2.80 mol Na(s) to grams by using the gram atomic mass:

$$\left(\frac{2.80 \text{ mol Na}(s)}{1} \right) \left(\frac{23.0 \text{ g Na}}{1 \text{ mol Na}} \right) = 64.4 \text{ g Na}(s)$$

6 **1.89g Al(s).** First, write out the reduction half of the reaction to determine how many moles of electrons you need to reduce the aluminum:

$$\text{Reduction half-reaction: } Al^{3+}(aq) + 3e^- \rightarrow Al(g)$$

This means you need 3 mol of electrons to reduce each mole of aluminum. Next, convert 45 minutes into seconds:

$$(45 \text{ min}) \left(\frac{60 \text{ s}}{1 \text{ min}} \right) = 2{,}700 \text{ s}$$

Multiply 2,700 seconds by the current of 8 A to get coulombs:

$$(2{,}700 \text{ s})(8 \text{ A}) = 21{,}600 \text{ C}$$

Divide this charge by 96,500 to give you the moles of electrons produced:

$$(21{,}600 \text{ C}) \left(\frac{1 \text{ mol e}^-}{96{,}500 \text{ C}} \right) = 0.22 \text{ mol e}^-$$

Divide this number by 3 to give you moles of aluminum:

$$(0.22 \text{ mol e}^-) \left(\frac{1 \text{ mol Al}}{3 \text{ mol e}^-} \right) = 0.07 \text{ mol Al}$$

Last, multiply the number of moles of aluminum by its molar mass (which is 26.98 g) to give you grams of Al(s):

$$(0.07 \text{ mol Al}) \left(\frac{26.98 \text{ g Al}}{1 \text{ mol Al}} \right) = 1.89 \text{ g Al}$$

Chapter 20

Doing Chemistry with Atomic Nuclei

・・

In This Chapter

▶ Decaying into alpha, beta, and gamma

▶ Living with half-lives

▶ Combining and splitting nuclei with fusion and fission

・・

Many elements in the periodic table exist in unstable versions called *radioisotopes* (see Chapter 3 for details). These radioisotopes decay into other (usually more stable) elements in a process called *radioactive decay.* Because the stability of these radioisotopes depends on the composition of their nuclei, radioactivity is considered a form of nuclear chemistry. Unsurprisingly, *nuclear chemistry* deals with nuclei and nuclear processes. Nuclear fusion, which fuels the sun, and nuclear fission, which fuels a nuclear bomb, are examples of nuclear chemistry because they deal with the joining or splitting of atomic nuclei. In this chapter, you find out about nuclear decay, rates of decay called *half-lives,* and the processes of fusion and fission.

Decaying Nuclei in Different Ways

Many radioisotopes exist, but not all radioisotopes are created equal. Radioisotopes break down through three separate decay processes (or decay *modes*): alpha decay, beta decay, and gamma decay. The following sections show you equations detailing each type of decay. **Note:** The symbols showing the isotope notation for each radioactive isotope are as follows: $^A_Z X$ or $^A_Z Y$, where

┃ ✔ X and Y represent the element symbols.

┃ ✔ Z represents atomic number.

┃ ✔ A represents the mass number (protons + neutrons).

Alpha decay

The first type of decay process, called *alpha decay,* involves emission of an alpha particle by the nucleus of an unstable atom. An alpha particle (*α* particle) is nothing more exotic than the nucleus of a helium atom, which is made of two protons and two neutrons. Emitting an alpha

particle decreases the atomic number of the daughter nucleus by 2 and decreases the mass number by 4. (See Chapter 3 for details about atomic numbers and mass numbers.) In general, then, a parent nucleus X decays into a daughter nucleus Y and an alpha particle.

$$_Z^A X \rightarrow _{Z-2}^{A-4} Y + _2^4 He$$

Beta decay

The second type of decay, called *beta decay* (*β* decay), comes in three forms, termed *beta-plus, beta-minus,* and *electron capture.* All three involve emission or capture of an electron or a *positron* (a particle with the tiny mass of an electron but with a positive charge), and all three also change the atomic number of the daughter atom.

✔ **Beta-plus:** In beta-plus decay, a proton in the nucleus decays into a neutron, a positron $\left(_{+1}^0 e\right)$, and a tiny, weakly interacting particle called a *neutrino* (*ν*). This decay decreases the atomic number by 1. The mass number, however, does not change. Both protons and neutrons are *nucleons* (particles in the nucleus), after all, each contributing 1 atomic mass unit. The general pattern of beta-plus decay is shown here:

$$_Z^A X \rightarrow _{Z-1}^A Y + _{+1}^0 e + \nu$$

✔ **Beta-minus:** Beta-minus decay essentially mirrors beta-plus decay. A neutron converts into a proton, emitting an electron and an *anti*neutrino (which has the same symbol as a neutrino except for the line on top). Particle and antiparticle pairs such as neutrinos and antineutrinos are a complicated physics topic, so we'll keep it basic here by saying that a neutrino and an antineutrino would annihilate one another if they ever touched, but they're otherwise very similar. Again, the mass number remains the same after decay because the number of nucleons remains the same. However, the atomic number increases by 1 because the number of protons increases by 1:

$$_Z^A X \rightarrow _{Z+1}^A Y + _{-1}^0 e + \bar{\nu}$$

✔ **Electron capture:** The final form of beta decay, electron capture, occurs when an inner electron — one in an orbital closest to the atomic nucleus — is "captured" by an atomic proton (see Chapter 4 for info on orbitals). By capturing the electron, the proton converts into a neutron and emits a neutrino. Here again, the atomic number decreases by 1:

$$_Z^A X + _{-1}^0 e \rightarrow _{Z-1}^A Y + \nu$$

Gamma decay

Gamma decay (*γ* decay) involves emission of a *gamma ray* (a high-energy form of light) by an excited nucleus. Although gamma decay changes neither the atomic number nor the mass number of the daughter nucleus, it often accompanies alpha or beta decay. Gamma decay allows the nucleus of a daughter atom to reach its lowest possible energy (most favorable) state. The general form of gamma decay is shown here, where $_Z^A X^*$ represents the excited state of the parent nucleus and the Greek letter gamma (*γ*) represents the gamma ray.

$$_Z^A X^* \rightarrow _Z^A Z + \gamma$$

When you think of dangerous radiation as portrayed in movies and on TV, you're thinking of gamma rays. They have an immense amount of energy and a very, very high frequency, which can kill livings cells easily. When you think of nuclear power, nuclear bombs, and other unsavory things you probably try to avoid on a daily basis, it's because you want to avoid their gamma rays. (Gamma rays also provide a wonderful benefit in cancer treatment because they can target, attack, and kill cancer cells.)

Q. If a parent nucleus has decayed into $^{218}_{86}$Rn through alpha decay, what was the parent nucleus, and what other particles were produced from the decay?

A. **The parent nucleus was radium, and an alpha particle was produced.** Alpha decay lowers the mass number by 4 and

the atomic number by 2, so the parent nucleus must have had a mass number of 222 and an atomic number of 88. Consulting your periodic table (check out an example in Chapter 4), you find that an atom with an atomic number of 88 is radium, so the parent nucleus must have been $^{222}_{88}$Rn.

---

1. Write out the complete formula for the alpha decay of a uranium nucleus with 238 nucleons.

Solve It

2. Sodium-22, a radioisotope of sodium with 22 nucleons, decays through electron capture. Write out the complete formula below, including all emitted particles.

Solve It

3. Classify the following reactions as alpha, beta, or gamma decay (specify the subtype if it's beta decay), and supply the missing particles.

a. $^{137}_{55}\text{Cs} \rightarrow \_\_ + ^{0}_{-1}\text{e} + \_\_$ Type:

b. $^{241}_{95}\text{Am} \rightarrow \_\_ + ^{4}_{2}\text{He}$ Type:

c. $^{60}_{28}\text{Ni}^* \rightarrow \_\_ + \_\_$ Type:

d. $^{11}_{6}\text{C} \rightarrow \_\_ + \_\_ + \nu$ Type:

Solve It

Measuring Rates of Decay: Half-Lives

The word *radioactive* sounds scary, but science and medicine are stuffed with useful, friendly applications for radioisotopes. Many of these applications are centered on the predictable decay rates of various radioisotopes. These predictable rates are characterized by half-lives. The *half-life* of a radioisotope is simply the amount of time it takes for exactly half of a sample of that isotope to decay into daughter nuclei. For example, if a scientist knows that a sample originally contained 42 mg of a certain radioisotope and measures 21 mg of that isotope in the sample four days later, then the half-life of that radioisotope is four days. The half-lives of radioisotopes range from seconds to billions of years.

Radioactive dating is the process scientists use to date samples based on the amount of radioisotope remaining. The most famous form of radioactive dating is carbon-14 dating, which has been used to date human remains and other organic artifacts. However, radioisotopes have also been used to date the Earth, the solar system, and even the universe.

Table 20-1 lists some of the more useful radioisotopes, along with their half-lives and decay modes (we discuss these modes earlier in this chapter).

Table 20-1	Common Radioisotopes, Half-Lives, and Decay Modes	
Radioisotope	*Half-Life*	*Decay Mode*
Carbon-14	5.73×10^3 years	beta
Iodine-131	8.0 days	beta, gamma
Potassium-40	1.25×10^9 years	beta, gamma
Radon-222	3.8 days	alpha
Thorium-234	24.1 days	beta, gamma
Uranium-238	4.46×10^9 years	alpha

To calculate the remaining amount of a radioisotope, use the following formula:

$$A = A_0(0.5)^{t/T}$$

where A_0 is the amount of the radioisotope that existed originally, A is the amount after the decay time, t is the amount of time the sample has had to decay, and T is the half-life. Be sure to pay attention to order of operations when you use this formula to ensure you get the correct answer.

Q. If a sample originally contained 1 g of thorium-234, how much of that isotope will the sample contain one year later?

A. **2.76×10^{-5} g.** Table 20-1 tells you that thorium-234 has a half-life of 24.1 days, so that's *T*. The time elapsed *(t)* is 365 days, and the original sample was 1 g *(A_0)*. Plugging these values into the half-life equation gives you

$$A = (1\text{ g})(0.5)^{365\text{ days}/24.1\text{ days}} = 2.76 \times 10^{-5}\text{ g}$$

To follow the order of operations, you must first perform the division found in the exponent, then perform the exponent calculation, and then multiply the answer by 1.

Notice that the units in the exponent cancel out, so you're left with grams in the end. These units make sense because you're measuring the mass of the sample remaining.

4. If you start with 0.5 g of potassium-40, how long will it take for the sample to decay to 0.1 g?

Solve It

5. If a 50 g sample of a radioactive element has decayed into 44.3 g after 1,000 years, what is the element's half-life? Based on Table 20-1, which radioisotope do you think you're dealing with?

Solve It

Making and Breaking Nuclei: Fusion and Fission

Fission and fusion differ from radioactive decay in that they generally require the nucleus of a parent atom to interact with an outside particle (some manmade isotopes have been known to undergo fission without bombardment — Fe-256, for example). Because the forces that hold atomic nuclei together are ridiculously powerful, the energy involved in splitting or joining two nuclei is tremendous. Here are the differences between fusion and fission:

✔ **Nuclear fusion:** In nature, nuclear fusion reactions occur only in the very center of stars like the sun, where extreme temperatures cause atoms of hydrogen, helium, and other light elements to smash together and join into one. This extremely energetic process, called *nuclear fusion,* is ultimately what causes the sun to shine. It almost always takes place with lighter elements like hydrogen and helium and is completely impossible with elements heavier than iron (Fe). Fusion also provides the outward pressure required to support the sun against gravitational collapse, an event that's even more dramatic than it sounds.

✔ **Nuclear fission:** *Nuclear fission,* the splitting of an atomic nucleus, doesn't occur in nature. Humans first harnessed the tremendous power of fission during the Manhattan Project, an intense, hush-hush effort by the United States that led to the development of the first atomic bomb in 1945. Fission has since been used for more-benign purposes in nuclear power plants. Nuclear power plants use a highly regulated process of fission to produce energy much more efficiently than is done in traditional, fossil fuel–burning power plants.

You can distinguish fission and fusion reactions from one another with a simple glance at products and reactants. If the reaction shows one large nucleus splitting into two smaller nuclei, then it's most certainly fission, whereas a reaction showing two small nuclei combining to make a single heavier nucleus is definitely fusion.

Q. What type of nuclear reaction — fusion or fission — would you expect plutonium-239 to undergo, and why?

A. Fission is the expected reaction. Plutonium has an exceptionally large number of nucleons and is therefore likely to undergo fission. Fusion is impossible in all elements heavier than iron, and plutonium is much, much heavier than iron.

6. What type of nuclear reaction is shown in the following equation? How do you know? Where might such a reaction take place?

$$^{235}_{92}U + ^{1}_{0}n \rightarrow ^{142}_{56}Ba + ^{91}_{36}Kr + 3^{1}_{0}n$$

Solve It

7. What type of nuclear reaction is shown in the following equation? How do you know? Where might such a reaction take place? There's something atypical about the two hydrogen reactants. What is it?

$$^{3}_{1}H + ^{2}_{1}H \rightarrow ^{4}_{2}He + ^{1}_{0}n$$

Solve It

Answers to Questions on Nuclear Chemistry

The following are the answers to the practice problems presented in this chapter.

1 $^{238}_{92}U \rightarrow {}^{234}_{90}Th + {}^{4}_{2}He$. Alpha decay results in the emission of a helium nucleus from the parent atom, which leaves the daughter atom with two fewer protons and a total of four fewer nucleons. In other words, the atomic number of the daughter atom is 2 fewer than the parent atom, and the mass number is 4 fewer. Because the proton number defines the identity of an atom, the element with two fewer protons than uranium must be thorium.

2 $^{22}_{11}Na + {}^{0}_{-1}e \rightarrow {}^{22}_{10}Ne + \nu$. Electron capture is a form of beta decay that results in the atomic number decreasing by 1 and the mass number remaining the same.

3 Here are the missing particles and the types of radioactive decay:

a. $^{137}_{55}Cs \rightarrow {}^{137}_{56}Ba + {}^{0}_{-1}e + \bar{\nu}$. **Type: beta-minus.** You can identify this reaction as beta-minus due to the emission of an electron ($^{0}_{-1}e$). This means that a neutron converts to a proton, which requires increasing the atomic number by 1. You also have to change the chemical symbol to reflect the element that's now present due to the change in atomic number.

b. $^{241}_{95}Am \rightarrow {}^{237}_{93}Np + {}^{4}_{2}He$. **Type: alpha.** This reaction is alpha decay due to the emission of an alpha particle, $^{4}_{2}He$. You simply need to adjust the atomic number and mass number to correspond to the loss of two neutrons and two protons. Thus, the mass number is reduced by 4, and the atomic number is reduced by 2. You then change the chemical symbol to reflect the element that's now present due to the change in atomic number.

c. $^{60}_{28}Ni^{*} \rightarrow {}^{60}_{28}Ni + \gamma$. **Type: gamma.** From the initial form of the equation, you can tell that an excited nucleus is present, so gamma decay will take place. To correctly write the equation, write the same isotope on the product side with a gamma ray being emitted.

d. $^{11}_{6}C \rightarrow {}^{11}_{5}B + {}^{0}_{+1}e + \nu$. **Type: beta-plus.** You see a neutrino in the products but no electron in the reactants, so this must be a beta-plus reaction. A neutron changes to a proton, so the mass number remains the same, but one more proton is added to the atomic number. This causes the element to change from carbon to boron.

4 **2.9×10^{9} years.** The problem tells you that $A = 0.1$ and $A_0 = 0.5$. Table 20-1 tells you that the half-life of potassium-40 is 1.25×10^{9} years. Plugging these values into the equation, you get

$$A = A_0(0.5)^{t/T}$$
$$0.1 = 0.5(0.5)^{(t/1.25 \times 10^{9} \text{ years})}$$

Divide both sides by 0.5:

$$0.2 = (0.5)^{(t/1.25 \times 10^{9} \text{ years})}$$

Take the natural log of both sides:

$$\ln(0.2) = \ln(0.5)^{(t/1.25 \times 10^{9} \text{ years})}$$

This step allows you to isolate the exponent by pulling it out in front on the right-hand side:

$$\ln(0.2) = \left(\frac{t}{1.25 \times 10^{9} \text{ years}} \right) \ln(0.5)$$

Compute the two natural logs:

$$-1.61 = \left(\frac{t}{1.25 \times 10^9 \text{ years}} \right)(-0.69)$$

Divide both sides by –0.69 to get

$$2.33 = \frac{t}{1.25 \times 10^9 \text{ years}}$$

Then multiply 2.32 by 1.25×10^9 to get $t = 2.9 \times 10^9$ years.

5 **5,714 years; carbon-14.** The problem tells you that $A_0 = 50$ g, $A = 44.3$ g, and $t = 1,000$ years. Use the same process as in Question 4 to isolate the exponent, this time solving for T instead of t.

$$A = A_0(0.5)^{t/T}$$
$$44.3 \text{ g} = (50 \text{ g})(0.5)^{(1,000 \text{ years}/T)}$$
$$0.886 = (0.5)^{(1,000 \text{ years}/T)}$$
$$\ln(0.886) = \ln(0.5)^{(1,000 \text{ years}/T)}$$
$$-0.121 = \left(\frac{1,000 \text{ years}}{T} \right)(-0.693)$$
$$0.175 = \frac{1,000 \text{ years}}{T}$$
$$T = 5,714 \text{ years}$$

The answer, 5,714 years, closely matches the half-life of carbon-14 in Table 20-1. (Try to keep the numbers in your calculator throughout each step of the process, and don't round until the end of the problem.)

6 This reaction is a fission reaction. It shows a heavy uranium nucleus being bombarded by a neutron and decaying into two lighter nuclei (barium and krypton). This is the very reaction that takes place in a nuclear reactor.

7 This reaction is a fusion reaction. It shows two light nuclei combining to form one heavy nucleus. This reaction fuels the sun. The two hydrogen reactants are atypical because they're rare isotopes of hydrogen, called *tritium* and *deuterium,* respectively.

Part V

The Part of Tens

web extras

Get tips on studying for a chemistry test in a free Part of Tens list at `www.dummies.com/extras/chemistrywb`.

In this part . . .

✔ Chemistry involves a great deal of formulas for calculating any number of different values. Here, we've brought together many of the most common formulas you'll encounter.

✔ Chemistry has a lot of rules, and sadly, most of those rules have exceptions. You can find a list of notable exceptions in this part.

Chapter 21

Ten Chemistry Formulas to Tattoo on Your Brain

In This Chapter

▶ Gathering important formulas for your memorizing pleasure

▶ Using handy shortcuts to make your calculations easier

*F*or some people, the mere presence of an equation is scary. Take consolation in this thought: Understanding an equation saves you brain space because equations pack a whole lot of explanation into one compact, little sentence.

For example, say you need to remember what decreasing temperature does to the volume of a gas when the pressure is held constant. Maybe you'll just remember. If not, you can try to reason out the relationship based on kinetic theory. That approach may not work for you, either. But if you simply remember the combined gas law, you can figure out the effect on volume simply by inspecting the equation. Not only that, but all other combinations of increasing or decreasing temperature, pressure, and volume fall instantly within your grasp. Equations can be your friends. Your friends have gathered to meet you in this chapter. Hang out for a while.

The Combined Gas Law

Using the following equation for the combined gas law, which we introduce in Chapter 11, you can calculate how a gas responds when one factor changes. Changing any one parameter (temperature, T; pressure, P; or volume, V) affects another parameter when the third parameter is held constant. To work the equation, first cancel out the constant parameter. Then plug in the values for your known parameters. Finally, solve for the unknown. (Remember, when using temperature, convert degrees Celsius to kelvins by adding 273 to any Celsius temperature.)

$$\frac{P_1 V_1}{T_1} = \frac{P_2 V_2}{T_2}$$

Values with a subscript 1 refer to initial states, and values with a subscript 2 refer to final states.

Dalton's Law of Partial Pressures

The following law, which we introduce in Chapter 11, is useful with mixtures of gases at constant volume and temperature. *Dalton's law of partial pressures* states that the total pressure of a mixture is simply the sum of the partial pressures of each of the individual gases. You can solve for the total pressure (P_{total}) or any individual partial pressure (P_1, P_2, and so on) as long as you know all the other pressures.

$$P_{total} = P_1 + P_2 + P_3 + \ldots + P_n$$

The Dilution Equation

The following formula works because the number of moles of a solute doesn't change when you dilute the solution. You can use the dilution equation to calculate an initial concentration or volume (M_1, V_1) or a final concentration or volume (M_2, V_2), as long as you know the other three values. Volumes don't have to be in particular units, as long as they're the same. See Chapter 12 for more information on dilutions.

$$M_1 V_1 = M_2 V_2$$

Rate Laws

Rate laws, which we cover in Chapter 14, relate reaction rates to the concentrations of reactants. Which rate law is appropriate depends on the kind of reaction involved:

- ✔ *Zero-order reactions* have rates that don't depend on reactant concentrations.

- ✔ *First-order reactions* have rates that depend on the concentration of one reactant.

- ✔ *Second-order reactions* have rates that depend on the concentrations of two reactants.

Here are representative reaction equations for zero-order, first-order, and second-order reactions (where the brackets stand for molar concentration):

Zero-order:	$A \rightarrow B$	Rate is independent of [A]
First-order:	$A \rightarrow B$	Rate depends on [A]
Second-order:	$A + A \rightarrow C$ $A + B \rightarrow C$	Rate depends on $[A]^2$ Rate depends on [A] and [B]

The rate laws for these reactions describe the rate at which Reactant A disappears over time:

$$\frac{-d[A]}{dt}$$

Note that $\frac{dx}{dt}$ simply stands for the rate at which some variable x is changing at any given moment in time, t. Rate laws are expressed in terms of k, the rate constant:

Zero-order rate law:	$\text{rate} = \dfrac{-d[A]}{dt} = k$
First-order rate law:	$\text{rate} = \dfrac{-d[A]}{dt} = k[A]$
Second-order rate law:	$\text{rate} = \dfrac{-d[A]}{dt} = k[A]^2 \ or \ \text{rate} = \dfrac{-d[A]}{dt} = k[A][B]$

The Equilibrium Constant

You can calculate the equilibrium constant for a reaction, K_{eq}, from the concentrations of reactants and products at equilibrium. In the following reaction, for example, A and B are reactants, C and D are products, and a, b, c, and d are *stoichiometric coefficients* (numbers showing mole multiples in a balanced equation):

$$aA + bB \leftrightarrow cC + dD$$

The equilibrium constant for the reaction is calculated as follows:

$$K_{eq} = \frac{[C]^c [D]^d}{[A]^a [B]^b}$$

Note that the general form [X] refers to the molar concentration of Reactant or Product X. For example, [0.7] = 0.7 mol/L.

Spontaneous reactions have K_{eq} values greater than 1. Nonspontaneous reactions have K_{eq} values between 0 and 1. The inverse of a K_{eq} value is the K_{eq} for the reverse reaction.

Check out Chapter 14 for details on the equilibrium constant.

Free Energy Change

Free energy, G, is the amount of energy available to do work in a system. Usually, the reactants and products of a chemical reaction possess different amounts of free energy, so the reaction proceeds with a change in energy between reactant and product states: $\Delta G = G_{product} - G_{reactant}$. The change in free energy arises from an interplay between the change in enthalpy, ΔH, and the change in entropy, ΔS:

$$\Delta G = \Delta H - T\Delta S$$

where T is temperature in kelvins.

Spontaneous reactions release energy and occur with a negative ΔG. Nonspontaneous reactions require an input of energy to proceed and occur with a positive ΔG.

The change in free energy associated with a reaction is related to the equilibrium constant for that reaction (see the preceding section for more about the equilibrium constant), so you can convert between ΔG and K_{eq}:

$$\Delta G = RT \ln K_{eq}$$

where R is the gas constant, T is temperature, and $\ln$ stands for natural log (that is, log base e). You'll find a button for ln on any respectable scientific calculator.

Flip to Chapter 14 for more information about free energy.

Constant-Pressure Calorimetry

Calorimetry is the measurement of heat changes that accompany a process (see Chapter 15 for details). The important values to know are heat (q), mass (m), specific heat capacity (C_p), and the change in temperature (ΔT). If you know any three of these values, you can calculate the fourth with this equation:

$$q = mC_p\Delta T$$

Be sure that your units of heat, mass, and temperature match those used in your specific heat capacity before attempting any calculations.

Hess's Law

The heat taken up or released by a chemical process is the same, regardless of whether the process occurs in one or several steps. So for a multistep reaction (at constant pressure), such as $A \rightarrow B \rightarrow C \rightarrow D$, the heats of the individual steps add up to the total heat for the reaction:

$$\Delta H_{A \rightarrow D} = \Delta H_{A \rightarrow B} + \Delta H_{B \rightarrow C} + \Delta H_{C \rightarrow D}$$

So you can calculate the overall change in heat or the change in heat for any given step, as long as you know all the other values. This formula is known as *Hess's law*.

Moreover, the reverse of any reaction occurs with the opposite change in heat:

$$\Delta H_{A \rightarrow D} = -\Delta H_{D \rightarrow A}$$

This means you can use known heat changes for reverse reactions simply by changing their signs. Flip to Chapter 15 for the full scoop on Hess's law.

pH, pOH, and K_W

As we explain in Chapter 16, pH and pOH are measurements of the acidity or basicity of aqueous solutions:

$$pH = -\log[H^+]$$

$$pOH = -\log[OH^-]$$

Pure water spontaneously self-ionizes to a small degree, leading to equal concentrations of hydrogen and hydroxide ions (1×10^{-7} mol/L, or 1×10^{-7} M):

$$H_2O(l) \leftrightarrow H^+(aq) + OH^-(aq)$$

(***Note:*** H+ exists in water as H_3O^+.) The product of these two concentrations is the ion-product constant for water, K_W:

$$K_W = [H^+][OH^-] = (10^{-7})(10^{-7}) = 10^{-14}$$

As a result, the pH and the pOH of pure water both equal 7. Acidic solutions have pH values less than 7 and pOH values greater than 7. Basic (alkaline) solutions have pH values greater than 7 and pOH values less than 7.

When you add an acid or base to an aqueous solution, the concentrations of hydrogen and hydroxide ions shift in proportion so that the following is always true:

$$pH + pOH = 14$$

K_a and K_b

As we explain in Chapter 16, the K_a and K_b are equilibrium constants that measure the tendency of weak acids and bases, respectively, to undergo ionization reactions in water. For the dissociation reaction of a weak acid, HA

$$HA \leftrightarrow H^+ + A^-$$

(also written $HA + H_2O \leftrightarrow A^- + H_3O^+$), the acid dissociation constant, K_a, is

$$K_a = \frac{[H^+][A^-]}{[HA]}$$

Note that H_3O^+ may be used in place of H^+.

For the dissociation reaction of a weak base, A^-

$$A^- + H_2O \leftrightarrow HA + OH^-$$

the base dissociation constant, K_b, is

$$K_b = \frac{[HA][OH^-]}{[A^-]}$$

Note that B^+ and BOH (base molecule, without or with OH^-) may be used in place of HA and A^- (acid molecule, with or without H^+).

Moreover, because $K_w = [H^+][OH^-]$, as we explain in the preceding section, it follows that

$$K_b = \frac{K_w}{K_a}$$

The stronger the acid or base, the larger the value of K_a or K_b.

It's often useful to consider pK_a or pK_b, the negative log of K_a or K_b:

$$pK_a = -\log K_a$$
$$pK_b = -\log K_b$$

The pK_a or pK_b is equivalent to the pH at which half the acid or base has undergone the dissociation reaction. This equivalence is reflected in the Henderson-Hasselbalch equation, which you can use to relate pH to the relative concentrations of a weak acid (HA) and its conjugate base (A^-):

$$pH = pK_a + \log\left(\frac{[A^-]}{[HA]}\right)$$

Flip to Chapter 17 for more information about the Henderson-Hasselbalch equation.

Chapter 22

Ten Annoying Exceptions to Chemistry Rules

In This Chapter

▶ Putting all the annoying exceptions into one corral

▶ Handling exceptions more easily with tips and tricks

*E*xceptions seem like nature's way of hedging its bets. As such, they're annoying — why can't nature just go ahead and *commit?* But nature knows nothing of our rules, so it certainly isn't going out of its way to annoy you. Nor is it going out of its way to make things easier for you. Either way, seeing as you have to deal with exceptions, we thought we'd corral many of them into this chapter so you can confront them more conveniently.

Hydrogen Isn't an Alkali Metal

In the field of psychology, Maslow's hierarchy of needs declares that people need a sense of belonging. It's a good thing hydrogen isn't a person, because it belongs nowhere on the periodic table. Although hydrogen is usually listed atop Group IA along with the alkali metals, it doesn't really fit. Sure, hydrogen can lose an electron to form a +1 cation, just like the alkali metals, but hydrogen can also gain an electron to form hydride, H^-, especially when bonding to metals. Furthermore, hydrogen doesn't have metallic properties, but it typically exists as the diatomic gas H_2.

These differences arise largely from the fact that hydrogen has only a single $1s$ orbital and lacks other, more interior orbitals that could shield the valence electrons from the positive charge of the nucleus. (See Chapter 4 for an introduction to orbitals.)

The Octet Rule Isn't Always an Option

An *octet*, as we explain in Chapter 4, is a full shell of eight valence electrons. The *octet rule* states that atoms bond with one another to acquire completely filled valence shells that contain eight electrons. It's a pretty good rule. Like most pretty good rules, it has exceptions:

✔ Atoms containing only $1s$ electrons simply don't have eight slots to fill, so hydrogen and helium obey the *duet* (two) *rule.* They're perfectly happy with only two electrons.

✔ Certain molecules contain an odd number of valence electrons. In these cases, full octets aren't an option. Like people born with an odd number of toes, these molecules may not be entirely happy with the situation, but they deal with it.

✔ Atoms often attempt to fill their valence shells by covalent bonding (see Chapter 5). Each covalent bond adds a shared electron to the shell. But covalent bonding usually requires an atom to donate an electron of its own for sharing within the bond. Some atoms run out of electrons to donate and therefore can't engage in enough covalent bonds to fill their shell octets. Boron trifluoride, BF_3, is a typical example. The central boron atom of this molecule can engage in only three B–F bonds and ends up with only six valence electrons. This boron is said to be *electron deficient.* You might speculate that the fluorine atoms could pitch in a bit and donate some more electrons to boron, but fluorine is highly electronegative and greedily holds fast to its own octets. C'est la vie.

✔ Some atoms take on more than a full octet's worth of electrons. These atoms are said to be *hypervalent* or *hypercoordinated.* The phosphorus of phosphorus pentachloride, PCl_5, is an example. These kinds of situations require an atom from Period (row) 3 or higher within the periodic table. The exact reasons for this restriction are still debated. Certainly, the larger atomic size of these atoms allows room to accommodate the bulk of all the binding partners that distribute around the central atom's valence shell. In some cases, even noble gases like xenon (Xe) form compounds. Because noble gases already have a filled valence shell, they automatically violate the octet rule.

Some Electron Configurations Ignore the Orbital Rules

Electrons fill orbitals from lowest energy to highest energy. This is true.

The progression of orbitals from lowest to highest energy is predicted by an Aufbau diagram. This isn't always true. Some atoms possess electron configurations that deviate from the standard rules for filling orbitals from the ground up. For Aufbau's sake, why?

Two conditions typically lead to exceptional electron configurations:

✔ Successive orbital energies must lie close together, as is the case with $3d$ and $4s$ orbitals.

✔ Shifting electrons between these energetically similar orbitals must result in a half-filled or fully filled set of identical orbitals, an energetically happy state of affairs.

Want a couple of examples? Strictly by the rules, chromium should have the following electron configuration:

$$[\text{Ar}]\,3d^4 4s^2$$

Because shifting a single electron from $4s$ to the energetically similar $3d$ level half-fills the $3d$ set, the actual configuration of chromium is

$$[\text{Ar}]\,3d^5 4s^1$$

For similar reasons, the configuration of copper is not the expected $[Ar]3d^9 4s^2$ but instead is $[Ar]3d^{10}4s^1$. Shifting a single electron from $4s$ to $3d$ fills the $3d$ set of orbitals. This shifting of electrons to violate the electron-filling rules is directly tied to an atom's need for stability. Other atoms that violate the electron-filling rules are copper, silver, and gold.

Flip to Chapter 4 for details on electron configurations and Aufbau diagrams.

One Partner in a Coordinate Covalent Bond Giveth Electrons; the Other Taketh

To form a covalent bond (as we explain in Chapter 5), each bonding partner contributes one electron to a two-electron bond, right? Not always.

In a *coordinate covalent bond,* one partner gives two electrons, and the other takes 'em. Coordinate covalent bonds are particularly common between transition metals (which are mostly listed as Group B elements on the periodic table) and partners that possess lone pairs of electrons. *Lone pairs* are pairs of nonbonding electrons within a single orbital, and transition metals have empty valence orbitals. So transition metals and lone-pair bearing molecules can engage in Lewis acid-base interactions (see Chapter 16). The lone-pair containing molecule acts as an electron donor (a Lewis base), giving both electrons to a bond with the metal, which acts as an electron acceptor (a Lewis acid).

The partners that bind to the metal are called *ligands,* and the resulting molecule is called a *coordination complex.* Coordination complexes are often intensely colored and can have properties that are quite different from those of the free metal.

All Hybridized Orbitals Are Created Equal

Different orbital types have grossly different shapes. Spherical s orbitals look nothing like lobed p orbitals, for example. So if the valence shell of an atom contains both s- and p-orbital electrons, you might expect those electrons to behave differently when it comes to things like bonding, right? Wrong. If you attempt to assume such a thing, valence bond theory politely taps you on the shoulder to remind you that valence shell electrons occupy hybridized orbitals. These hybridized orbitals (as in sp^3, sp^2, and sp orbitals) reflect a mixture of the properties of the orbitals that make them up, and each of the orbitals is equivalent to the others in the valence shell.

Although this phenomenon represents an exception to the rules, it's somewhat less annoying than other exceptions because hybridization allows for the nicely symmetrical orbital geometries of actual atoms within actual molecules. VSEPR theory presently clears its throat to point out that the negative charge of the electrons within the hybridized orbitals causes those equivalent orbitals to spread as far apart as possible from one another. As a result, the geometry of sp^3-hybridized methane (CH_4), for example, is beautifully tetrahedral.

Check out Chapter 5 for the details on VSEPR theory and hybridization.

Use Caution When Naming Compounds with Transition Metals

The thing about transition metals is that the same transition metal can form cations with different charges. Differently charged metal cations need different names so chemists don't get any more confused than they already are. These days, you indicate these differences by using Roman numerals within parentheses to denote the positive charge of the metal ion. However, an older method adds the suffix *-ous* or *-ic* to indicate the cation with the smaller or larger charge, respectively. For example:

Cu^+ = copper (I) ion or cuprous ion

Cu^{2+} = copper (II) ion or cupric ion

Metal cations team up with nonmetal anions to form ionic compounds. What's more, the ratio of cations to anions within each formula unit depends on the charge assumed by the fickle transition metal. The formula unit as a whole must be electrically neutral. The rules you follow to name an ionic compound must accommodate the whims of transition metals. The system of Roman numerals or suffixes applies in such situations:

$CuCl$ = copper (I) chloride or cuprous chloride

$CuCl_2$ = copper (II) chloride or cupric chloride

Chapter 6 has the full scoop on the modern system of naming ionic and other types of compounds.

You Must Memorize Polyatomic Ions

Sorry, it's true. Not only are polyatomic ions annoying because you have to memorize them, but they pop up everywhere. If you don't memorize the polyatomic ions, you'll waste time trying to figure out weird (and incorrect) covalent bonding arrangements when what you're really dealing with is a straightforward ionic compound. Here are the common polyatomic ions in Table 22-1 (see Chapter 6 for more information on these ions).

Table 22-1	Common Polyatomic Ions		
−1 Charge	*−2 Charge*	*−3 Charge*	*+1 Charge*
Dihydrogen phosphate $(H_2PO_4^-)$	Hydrogen phosphate (HPO_4^{2-})	Phosphite (PO_3^{3-})	Ammonium (NH_4^+)
Acetate $(C_2H_3O_2^-)$	Oxalate $(C_2O_4^{2-})$	Phosphate (PO_4^{3-})	
Hydrogen sulfite (HSO_3^-)	Sulfite (SO_3^{2-})		
Hydrogen sulfate (HSO_4^-)	Sulfate (SO_4^{2-})		
Hydrogen carbonate (HCO_3^-)	Carbonate (CO_3^{2-})		

−1 Charge	−2 Charge	−3 Charge	+1 Charge
Nitrite (NO_2^-)	Chromate (CrO_4^{2-})		
Nitrate (NO_3^-)	Dichromate ($Cr_2O_7^{2-}$)		
Cyanide (CN^-)	Silicate (SiO_3^{2-})		
Hydroxide (OH^-)			
Permanganate (MnO_4^-)			
Hypochlorite (ClO^-)			
Chlorite (ClO_2^-)			
Chlorate (ClO_3^-)			
Perchlorate (ClO_4^-)			

Liquid Water Is Denser than Ice

Kinetic molecular theory, which we cover in Chapter 10, predicts that adding heat to a collection of particles increases the volume occupied by those particles. Heat-induced changes in volume are particularly evident at phase changes, so liquids tend to be less dense than their solid counterparts. Weird water throws a wet monkey wrench into the works. Because of H_2O's ideal hydrogen-bonding geometry, the lattice geometry of solid water (ice) is very "open," with large, empty spaces at the center of a hexagonal ring of water molecules. These empty spaces lead to a lower density of solid water relative to liquid water, so ice floats in water. Although annoying, this watery exception is quite important for biology.

No Gas Is Truly Ideal

No matter what your misty-eyed grandparents tell you, there were never halcyon Days of Old when all the gases were ideal. To be perfectly frank, not a single gas is really, truly ideal. Some gases just approach the ideal more closely than others. At very high pressures, even gases that normally behave close to the ideal cease to follow the ideal gas laws that we talk about in Chapter 11.

When gases deviate from the ideal, we call them *real gases*. Real gases have properties that are significantly shaped by the volumes of the gas particles and/or by interparticle forces. To account for these nonideal factors, chemists use the van der Waals equation. Compared to the ideal gas equation, $PV = nRT$, the following van der Waals equation includes two extra variables, a and b. The variable a corrects for effects due to particle volume. The variable b corrects for interparticle forces. The van der Waals equation is appropriate for gases at very high pressure, in low-temperature conditions, and when gas particles have strong mutual attraction or repulsion.

$$\left(P + \frac{n^2 a}{V^2}\right)(V - nb) = nRT$$

Common Names for Organic Compounds Hearken Back to the Old Days

Serious study of chemistry predates modern systematic methods for naming compounds. As a result, chemists persistently address a large number of common compounds, especially organic compounds, by older, "trivial" names. This practice won't change anytime soon. A cynical take on the situation is to observe that progress occurs one funeral at a time. A less cynical approach involves serenely accepting that which you cannot change and getting familiar with these old-fashioned names. Table 22-2 lists some important ones.

Table 22-2	Common Names for Organic Compounds	
Formula	*Systematic Name*	*Common Name*
CH_3CO_2H	Ethanoic acid	Acetic acid
CH_3COCH_3	Propanone	Acetone
C_2H_2	Ethyne	Acetylene
$CHCl_3$	Trichloromethane	Chloroform
C_2H_4	Ethene	Ethylene
H_2CO	Methanal	Formaldehyde
CH_2O_2	Methanoic acid	Formic acid
C_3H_8O	Propan-2-ol	Isopropanol

Index

About the Authors

Peter Mikulecky grew up in Milwaukee, an area of Wisconsin unique for its high human-to-cow ratio. After a breezy four-year tour in the Army, Peter earned a bachelor of science degree in biochemistry and molecular biology from the University of Wisconsin–Eau Claire and a PhD in biological chemistry from Indiana University. With science seething in his DNA, he sought to infect others with a sense of molecular wonderment. Having taught, tutored, and mentored in classroom and laboratory environments, Peter was happy to find a home at Fusion Learning Center and Fusion Academy. There, he enjoys convincing students that biology and chemistry are, in fact, fascinating journeys, not entirely designed to inflict pain on hapless teenagers. His military training occasionally aids him in this effort.

Chris Hren is a high school chemistry teacher. He has been happily married for a while now and loves spending time with his family. Chris is a proud graduate of Michigan State University and spends every Saturday in the fall watching his Spartans. Chris has coached football and track and is a pretty big sports fan in general. His family is lucky enough to have a small cottage on a lake in Northern Michigan, where he enjoys many outdoor activities. Chris loves to read, especially science fiction books.

Dedication

Peter Mikulecky: I would like to dedicate this book to my family and friends who supported me during the writing process. Also, to all my students who motivate me to be a better teacher by pushing me to find unique and fresh ways to reach them.

Chris Hren: I would like to dedicate this book to my wife. Without her support, none of this would have been possible. I love her dearly.

Authors' Acknowledgments

Peter Mikulecky: Thanks to Bill Gladstone from Waterside Productions for being an amazing agent and friend. Thanks to Georgette Beatty, our project editor, for her clear feedback and support. A special shout-out to our acquisitions editor, Lindsay Lefevere, who, for reasons unclear, seems to keep wanting to work with us.

Chris Hren: Thanks to my wife and family again. A special thanks goes out to Georgette Beatty and Danielle Voirol for being such wonderful editors. Thank you as well to my students.

Publisher's Acknowledgments

Executive Editor: Lindsay Sandman Lefevere

Senior Project Editor: Georgette Beatty

Copy Editor: Danielle Voirol

Technical Editors: Jason Dunham, Patti Smykal

Project Coordinator: Emily Benford

Cover Image: ©iStock.com/Eraxion

31901055827085